INTEGRATED LAND USE PLANNING FOR SUSTAINABLE AGRICULTURE AND RURAL DEVELOPMENT

INTEGRATED LAND USE PLANNING FOR SUSTAINABLE AGRICULTURE AND RURAL DEVELOPMENT

Edited by
M. V. Rao, PhD, V. Suresh Babu, PhD,
K. Suman Chandra, PhD, and G. Ravindra Chary, PhD

Apple Academic Press Inc. | Apple Academic Press Inc.
3333 Mistwell Crescent | 9 Spinnaker Way
Oakville, ON L6L 0A2 | Waretown, NJ 08758
Canada | USA

© 2016 by Apple Academic Press, Inc.

First issued in paperback 2021

Exclusive worldwide distribution by CRC Press, a member of Taylor & Francis Group

No claim to original U.S. Government works

ISBN-13: 978-1-77463-379-3 (pbk)
ISBN-13: 978-1-77188-104-3 (hbk)

Library and Archives Canada Cataloguing in Publication

Integrated land use planning for sustainable agriculture and rural development / edited by M.V. Rao, PhD, V. Suresh Babu, PhD, K. Suman Chandra, PhD, and G. Ravindra Chary, PhD.

Includes bibliographical references and index.
ISBN 978-1-77188-104-3 (bound)
1. Land use, Rural--India--Planning. 2. Sustainable agriculture--India. 3. Rural development--India. 4. Geographic information systems--India. 5. India--Rural conditions. I. Chandra, K. Suman, author, editor II. Suresh Babu, S. V., author, editor III. Rao, M. V., 1962-, author, editor IV. Ravindra Chary, G., author, editor

HD876.5.I57 2015   333.73'130954   C2015-903757-3

Library of Congress Cataloging-in-Publication Data

Rao, M. V., 1962- author.
Integrated land use planning for sustainable agriculture and rural development / authors: M. V. Rao, PhD; V. Suresh Babu, PhD; K. Suman Chandra, PhD; and G. Ravindra Chary, PhD. -- 1st ed.

pages cm
Includes bibliographical references and index.
ISBN 978-1-77188-104-3 (alk. paper)
1. Land use--India--Planning. 2. Sustainable agriculture--India. 3. Rural development--India. I. Suresh Babu, V., author. II. Chandra, K. Suman, author. III. Ravindra Chary, G., author. IV. Title.

HD876.5.R36 2015   333.76'130954--dc23   2015019039

Apple Academic Press also publishes its books in a variety of electronic formats. Some content that appears in print may not be available in electronic format. For information about Apple Academic Press products, visit our website at **www.appleacademicpress.com** and the CRC Press website at **www.crcpress.com**

ABOUT THE EDITORS

M. V. Rao, PhD

M. V. Rao, PhD, is presently working as Director General of the National Institute of Rural Development and also as Chief Executive of the National Fisheries Development Board, Ministry of Fisheries, Government of India. He obtained his doctoral degree in Development Administration from Sambalpur University, Sambalpur, Odisha, India. He belongs to the 1988 batch of the Indian Administrative Service and served the state and the nation in different capacities for the uplifting of rural communities. As District Magistrate of the largest district in India, with a population of one million, and as Commissioner of Rural Development, Government of West Bengal, he has helped millions of small and marginal resource-poor farmers in the alleviating their poverty and enhancing their standard of living. Publications in which he has been involved include *Community Health in Community Hands, Development for the People: Innovations in Administration, Persistence and Change in Tribal India–Saga of Tribal People of West Midnapore,* and *Panchayats and Building of Model Villages.*

V. Suresh Babu, PhD

V. Suresh Babu, PhD, is currently Associate Professor at the National Institute of Rural Development, Hyderabad, India. He earned a doctoral degree in agronomy from the University of Agricultural Sciences, Bangalore, Karnataka. He has worked at the Regional Centre of the National Afforestation and Eco-Development Board (NAEB) in Bangalore. His published books include *Indigenous and Traditional Knowledge for Promotion of Sustainable Agriculture, Agrarian Crisis in India, The Way Out, Comparative Assessment of Success or Failure of JFM and Its Impact on Natural Forest Cover in Andhra Pradesh, Karnataka and Tamil Nadu, Role of Tree Shelterbelts in Coastal Areas in Mitigating Harsh Effects of Cyclones in Andhra Pradesh and Tamil Nadu,* and *Risk, Vulnerability and Coping Mechanisms in Rainfed Agriculture-A Study in Three States.* He currently teaches agriculture courses on integrated farming system, bio-inputs, agro-forestry, watersheds, mangroves restoration, and joint forest management, etc.

K. Suman Chandra, PhD

K. Suman Chandra, PhD, is currently Professor and Head of the Centre for Agrarian Studies at the National Institute of Rural Development in Hyderabad, India. He obtained a doctoral degree in sociology from Andhra University, Visakhapatnam, India. His areas of specialization include agrarian reforms in general and tenancy reforms in particular. His published books include *Agrarian Crisis in India, The Way Out; Indigenous and Traditional Knowledge for Promotion of Sustainable Agriculture; Proceedings of the National Workshop on Accelerating Rural Development and Strengthening Local Self Governance; Evaluation of National Child Labour Projects in AP,* and *Farmers Suicides in AP and Karnataka.* He currently teaches courses in the field of agrarian reforms, tenancy reforms, wage and income mobility in Indian labor market, and disaster management including drought mitigation and coping strategies.

G. Ravindra Chary, PhD

G. Ravindra Chary, PhD, is currently working as a Principle Scientist in the All India Coordinated Research Project for Dryland Agriculture (AICRP-DA) of the Central Research Institute for Dryland Agriculture (CRIDA), Hyderabad, India. He received a doctoral degree in agronomy and has served in different capacities in the Indian Council of Agriculture Research (ICAR) institutions. He had worked at the National Bureau for Soil Survey and Land Use Planning (NBSSLUP), Nagpur, for a period of 10 years. He has published 20 books on such topics as land use planning, drought management, rainfed farming—a compendium of doable technologies, real-time contingency planning, etc. Some of his publications include *Rainfed Farming–A Compendium of Doable Technologies, Adaptation and Mitigation Strategies for Climate Resilient Agriculture, District Level Contingency Planning for Weather Aberrations in Himachal Pradesh, Contingency Crop Planning for 100 Districts in Peninsular India,* and *District Level Contingency Plans for Weather Aberrations in India.* He teaches courses on land use planning, rainfed farming, drought management, rainfed farming systems, contingency planning, agronomic management practices including soil and moisture conservation, and crop planning and cropping systems.

CONTENTS

LIST OF CONTRIBUTORS

V. Suresh Babu
Centre for Agrarian Studies and Disaster Mitigation, National Institute of Rural Development & Panchayati Raj, Hyderabad – 500 030, Andhra Pradesh, India

Tapas Bhattacharyya
Division of Soil Resource Studies, National Bureau Soil Survey and Land Use Planning Amravati Road, Nagpur 440 033, Maharashtra, India

Bhuvaneshwari
Cochin Special Economic Zone, UL Cyber Park Project Office, Nellikode Village, Koshikode – 673 016, Kerala, India

K. Suman Chandra
Centre for Agrarian Studies and Disaster Mitigation, National Institute of Rural Development & Panchayati Raj, Hyderabad – 500 030, Andhra Pradesh, India

G. Ravindra Chary
Central Research Institute for Dryland Agriculture, Santoshnagar, Saidabad P.O., Hyderabad – 500 059, Andhra Pradesh, India

S. Chatterji
Division of Land Use Planning, National Bureau Soil Survey and Land Use Planning Amravati Road, Nagpur 440 033, Maharashtra, India

Arun Chaturvedi
Division of Land Use Planning, National Bureau Soil Survey and Land Use Planning Amravati Road, Nagpur 440 033, Maharashtra, India

T. Phanindra Kumar
Centre for Geomatic Application in Rural Development, National Institute of Rural Development & Panchayati Raj, Hyderabad – 500 030, Andhra Pradesh, India

M. J. Mercy Kutty
College of Agriculture, Padannakkadi, Kasaragod, Kerala, India

A. V. Meera Manjushah
Regional Agricultural Research Station, Pilicode, Kerala Agricultural University, Pilicode PO, Kasaragod, Kerala, India

M. Moni
National Informatics Centre, New Delhi, India

A. Natrajan
Regional Centre, National Bureau Soil Survey and Land Use Planning, Hebbal, Bangalore – 560 024, Karnataka, India

N. S. R. Prasad
Centre for Geomatic Application in Rural Development, National Institute of Rural Development & Panchayati Raj, Hyderabad – 500 030, Andhra Pradesh, India

A. Prema
College of Horticulture, Kerala Agricultural University, KAU – P.O., Thrissur – 680 656, Kerala, India

V. Ramamurthy
Regional Centre, National Bureau Soil Survey and Land Use Planning, Hebbal, Bangalore – 560 024, Karnataka, India

M. V. Rao
Director General, National Institute of Rural Development & Panchayati Raj Hyderabad – 500 030, Andhra Pradesh, India

P. Keshava Rao
Centre for Geomatic Application in Rural Development, National Institute of Rural Development & Panchayati Raj, Hyderabad – 500 030, Andhra Pradesh, India

V. Madhava Rao
Centre for Geomatic Application in Rural Development, National Institute of Rural Development & Panchayati Raj, Hyderabad – 500 030, Andhra Pradesh, India

V. U. M. Rao
Central Research Institute for Dryland Agriculture, Santoshnagar, Saidabad P.O., Hyderabad - 500 059, Andhra Pradesh, India

G. P. Obi Reddy
GIS Section, National Bureau Soil Survey and Land Use Planning Amravati Road, Nagpur 440 033, Maharashtra, India

A. K. Sahoo
National Bureau Soil Survey and Land Use Planning Block-DK, Sector II, Salt Lake, Kolkata 700 091, West Bengal, India

Dipak Sarkar
National Bureau Soil Survey and Land Use Planning Amravati Road, Shankarnagar, Nagpur – 440 010, Maharashtra, India

Mridula Singh
State Land Use Board, Department Of Planning, 5th Floor, Yojna Bhawan, Lucknow – 226 001 Uttar Pradesh, India

Rajeev Srivastava
Division of RSA, National Bureau Soil Survey and Land Use Planning Amravati Road, Nagpur 440 033, Maharashtra, India

M. Velayutham
6A, Gardenic Apartment, 275, Kilpank Garden Road, Kilpank, Chennai, Tamil Nadu, India

LIST OF ABBREVIATIONS

AESRs	Agro Ecological Sub Regions
AEU	Agro-Ecological Units
AIBP	Accelerated Irrigation Benefits Program
AICRPDA	All India Coordinated Research Project for Dryland Agriculture
AMMP	Agriculture Mission Mode Project
APIB	Agro Climatic Planning and Information Bank
APMC	Agriculture Produce Marketing Committees
ARWSP	Accelerated Rural Water Supply Program
AWC	Available Water Capacity
AWSs	Automatic Weather Stations
BRGF	Backward Region Grant Fund
BVF	Bhoovigyan Vikas Foundation
CADP	Command Area Development Program
CCD	Convention to Combat Desertification
CDM	Clean Development Mechanism
CGIAR	Consultative Group for International Agricultural Research
CLDP	Comprehensive Land Development Program
COAG	Committee on Agriculture
CSCs	Common Service Centers
CSD	UN Commission on Sustainable Development
DAC	Department of Agriculture and Cooperation
DBMS	Database Management System
DDP	Desert Development Program
DEM	Digital Elevation Models
DIP	Digital Image Processing
DPAP	Drought Prone Areas Program
DRS	Diffused Reflectance Spectroscopy
ECe	Electrical Conductivity
EPI	Earth Care Policy Institute
FSI	Food grains Security Index
GCA	Gross cropped area
GeoSIS	Geo-referencing soil information system
GHG	Greenhouse gas emission
GIS	Geographic Information System

GLASOD	Global Assessment of Soil Degradation
GNP	Gross National Product
GPS	Global Positioning System
HADP	Hill Area Development Program
ICT	Information and Communication Technology
IUCN	International Union for Conservation of Nature and Natural Resources
IWDP	Integrated Waste Land Development Program
IWMI	International Water Management Institute
IWMP	Integrated Watershed Management Program
KIF	Key Informant Farmers
LGP	Length of Growing Period
LMU	Land Management Units
LRIS	Land Resource Information System
LUP	Land Use Planning
MGNREGS	Mahatma Gandhi National Rural Employment Guarantee Scheme
MUDA	Mysore Urban Development Authority
NATP	National Agricultural Technology Project
NBSS	National Bureau of Soil Survey
NBSSLUP	National Bureau of Soil Survey and Land Use Planning
NCA	National Commission on Agriculture
NDVI	Normalized Difference Vegetation Index
NFP	National Forest Policy
NMSA	National Mission on Sustainable Agriculture
NREGS	National Rural Employment Guarantee Scheme
NRM	Natural Resource Management
NRSA	National Remote Sensing Agency
NUG	National University of Geography
OC	Organic Carbon
RAEZs	Rainfed Agroeconomic Zones
RDBMS	Relational database management system
RKVY	Rashtriya Krishi Vikas Yojana
RS	Remote Sensing
SARD	Sustainable Agriculture and Rural Development
SIS	Soil information system
SISIGP	Soil Information System for the Indo-Gangetic Plains
SOC	Soil Organic Carbon
SOTER	Soil and Terrain Digital Database
SPWD	Society for Promotion of Wasteland Developments
SCUs	Soil Conservation Units
SQUs	Soil Quality Units

SRTM	Shuttle Radar Topographic Mission
SWOT	Strengths, Weaknesses, Opportunities and Threats
TFP	Total Factor Productivity
UNFCC	UN Framework Convention on Climate Change
VCB	Vented Cross Bar
VNIR	Visble Near-Infrared
WAN	Wide Area Network
WSSD	World Summit on Sustainable Development
WUAs	Water Users' Associations

PREFACE

The increasing human population in developing countries is putting pressure on their finite land resources and causing land degradation. The majority of households depend on land and other natural resources for fulfilling their immediate needs and achieving their long-term livelihood ambitions. In developing countries like India, land degradation has been exacerbated in the absence of effective land use planning, leading to overexploitation of land resources. As a sequel, misery for large segments of local population and destruction of valuable bio-diversity is increasing by many folds.

An integrated approach to planning the use and management of land resources entails the involvement of all stakeholders in the process of decision making on the future of the land and the identification and evaluation of all biophysical and socio-economic attributes of land units. This requires the identification and establishment of a use or non-use of each land unit that is technically appropriate, economically viable, socially acceptable and environmentally non-degrading. Current land use issues, which require a resolution formulated with the aid of holistic approach in the rural areas, are frequently derived from environmental versus developmental conflicts.

To address the above issues a platform was created at the National Institute of Rural Development & Panchayati Raj (NIRD & PR), Ministry of Rural Development, Hyderabad, in technical collaboration with the National Bureau for Soil Survey and Land Use Planning (NBSSLUP), Nagpur and Central Research Institute for Dryland Agriculture (CRIDA), Hyderabad, during June 19–21, 2012. This publication is the outcome of a compilation of research papers presented by the researchers addressing the thematic areas of Theme I: Soil and Land Resource Information and Land Degradation Assessment; Theme II: Application of Geo-informatics and GIS in Land Use Planning; Theme III: Planning for Sustainable Agriculture; Land Use in Different Agro-ecosystems; and Theme IV: Integrated Land Use Planning and Institutional Arrangements. Over all twenty-one papers are presented in this volume, considering the location-specific agro-climatic conditions and zones.

We are highly thankful to Dr. M. Velayutham, former Deputy Director General (ICAR), and Dr. Dipak Sarkar, Director, NBSSLUP, Nagpur, for their guidance and support. As editors, we would like to thank all the authors for their efforts and cooperation in bringing out this edited volume. We express our thanks to Mrs. Padmashree; Mrs. K. Jayasree, Project Associate; Dr. P. Anuradha, Project Associate and Mr. Maharana, Project Associate; for their editorial assistance. We thank all the staff of NIRD who helped directly or indirectly in organizing this workshop successfully. We firmly believe that this publication will be highly useful, in one way or other, for researchers, academicians, extension workers, policymakers, planners, officials in land use institutions/agencies and students who are working on similar agro-climatic conditions across the globe.

ABOUT THE VOLUME

Land represents an important resource for the economic life of a majority of people in the world. The way people handle and use land resource is decisive for their social and economic well-being as well as for the sustained quality of land resources. Under the rainfed conditions in arid and, semi-arid regions, the land use is diversified into multiple crops under multi-tiered systems both spatially and temporally driven by economics. The cropping in irrigated and coastal agriculture are driven by the period of waterlogging while temperate agriculture is addressed to available cropping period. Thus the scopes of crops expand on a landscape depending on the location of field, but needs to coexist on a plane utilizing the endowed natural resources creating biocorridors and ways of thinking how land to be used. Henceforth come into existence many practices with experience of farmers.

In Asia, poverty has mainly been a rural phenomenon, and nearly three-fourth of the poor live in rural areas, with large majority of them dependent on natural resources for employment and income. South Asia, which had a poverty incidence of 43 percent, contributed about 40 percent of the world's poor. The way people handle and use land resource is decisive for their social and economic well-being as well as for the sustained quality of land resources. Land use planning is also integral to water resources development and management for agriculture, industry, drinking water, power generation, etc. The "National Land Use Policy guideline and action points" were prepared by the Government of India, Ministry of Agriculture after intensive deliberations. In the said policy, framing of suitable legislation and its sincere enforcement were stressed by imposing penalties, of violation thereof. The said policy guidelines were placed before the National Land Use and Waste Land Development Council under the chairmanship of the Prime Minister and were agreed to in principle. The land use plans to uplift rural poor and sustain growth are, although much talked about, left little in practice. The legislative support is negligible for obvious reasons of economics of the stakeholders. Under the existing contra-indications, the topic is forever a green futuristic subject, especially with the present day climate change.

The Government of India spends every year nearly Rs one lakh crore on various agricultural development programs, soil and water conservation being the most prominent activity. Various schemes and programs were initiated by central and state governments under the aegis of a Planning Commission and tried to address either directly or indirectly the LUP (land use planning) in sectorial manner. Land-based rural developmental activities were being implemented in a piece-meal, top-down manner, without consideration of land resources. In India, two major institutions dealing this subject directly—the National Institute of Rural Development (NIRD), Hyderabad and the National Bureau of Soil Survey and Land Use Planning, Nagpur—joined hands to review the state of art and discuss various avenues to overcome the hindrances. A workshop on "Integrated Land Use Planning for Sustainable Agriculture and Rural Development" was held in NIRD, on June 18–20, 2012. In a wide dimensioned theme seminar like "Land Use Planning for Sustainable Agriculture and Rural Development," it is customary to have wide-ranging discussions. The twenty-one contributions in four themes and ten working group discussions brought to the forefront the state of practice of land use planning in the country, governmental efforts like a Central Ministry for Earth Sciences, research, synergies to be utilized, and others among many. The backdrop was climate change, food security and self-reliance, use and reliability of information technology tools, amendments for legal framework, mobilization, capacity building, knowledge enhancement, environment, and international pressures to adopt appropriate earth care policies, etc. This workshop addressed the issues of:

What are the sources, scales and quality of land resources and land use data for efficient planning?

- Which are the hot spots in the agro-ecological systems that are degraded? How to rejuvenate such areas in terms of quality?
- What should be the unit of planning for holistic approach to conserve and efficient use of natural resources at different administrative levels/domains?
- What is the role of state land use board, district planning committees and its implications on LUP?
- How to integrate various programs and schemes implemented by central and state to address SARD (Sustainable agriculture and rural development)?
- What is the need, role and the capacities of the Gram Panchayat members for holistic planning and enhance the manpower at GP level for efficient planning?
- What are the technologies, new policies and support systems required to address SARD?
- What are the legal issues relevant to integrate LUP?
- What should be role of community in planning for poverty alleviation and integrated planning for SARD and support NRM based livelihoods?
- Constraints in preparing GIS-based planning and at what level GIS based planning is accurate and possible for integrated land use planning for SARD?

In brief, the proceedings can be summarized as follows:

Various agro-ecosystems covered in the workshop are irrigated, rainfed, coastal, semi-arid, arid (drylands), and hill and mountain (temperate) regions. Any development in the agro-ecosystems is sustainable only when an optimal and efficient plan to utilize natural resources is firmly in place. This implies that policy discussion and development planning need to be based on a sound understanding of these dynamics. Land utilization is a complicated issue and to some extent represents the man-environment relationship. Conflict over land use is inevitable. With high population growth, endemic poverty and weak existing institutional capacity for land management, a scientific and strategically important land use policy is must for sustainable development. In this context it is widely perceived that any state can only prosper and have sustainable growth pattern if it utilizes its natural resources properly and judiciously. Otherwise even with the most productive soils and abundant water, economic poverty and backwardness will prevail. In the backdrop of critical issues, that is, climate change, food security and self-reliance, use and reliability of information technology tools, available and proposed legal framework, mobilization, capacity building and knowledge enhancement related to land use planning it is a challenge and uphill task for policy makers and planners to recommend an acceptable and implementable land use policy framework. The Constitution of India also enables the Centre and the States to enact laws to carry out the duties of preservation, afforestation and conservation of natural resources. Environmental conservation has been an integral part of the Indian ethos and is reflected in the Constitution. The state of practice of land use planning in the country, synergies to be utilized, government efforts, and others are discussed in the contest of future food security.

This volume will be a ready reckoner on various aspects to all those interested in land use subject, including scientists, research scholars and policymakers, including commercial and non-commercial stakeholders.

AN ESSENCE OF THE WORKSHOP ON "INTEGRATED LAND USE PLANNING FOR SUSTAINABLE AGRICULTURE AND RURAL DEVELOPMENT"

K. P. R. VITTAL

Former Director, National Institute of Abiotic Stress Management (A Deemed to be University Under ICAR, MoA, GoI), Baramati, MS, India

with

M. V. RAO, V. SURESH BABU, and K. SUMAN CHANDRA

Director General, Associate Professor, Professor and Head, Centre for Agrarian Studies, National Institute of Rural Development and Panchayati Raj (NIRD & PR), Hyderabad – 500 030, India

G. RAVINDRA CHARY

Principal Scientist, Central Research Institute for Dryland Agriculture, CRIDA, Hyderabad – 500 059, India

ORDER OF THE WORKSHOP

Four objectives were set for the Workshop on "Integrated Land Use Planning for Sustainable Agriculture and Rural Development." Under each of the objectives the status was analyzed and need of the hour was identified. These needs were considered by ten separate sub-groups and recommendations were made along with the identification of appropriate agencies. An essence follows:

1 OBJECTIVE

To Review the Current Land Use Planning Practices.

1.1 STATUS

The review comprises of status on Land Degradation, State-of-Practice of Land Use and Mapping.

1.1.1 LAND DEGRADATION

Agriculture is still the primary sector in Indian economy. At the biophysical level, the negative impacts have led to reduction in Total Factor Productivity (TFP), soil organic carbon depletion, increase in soil salinity, imbalances in water regimes, and pollution due to indiscriminate application of fertilizers and pesticides. While, Green revolution's food production increased from about 5.1 mt in 1950 to over 205 mt now, the higher economic growth and population pressure demands about 260 million tons of food grains by 2030. At the biophysical level, the negative impacts have led to reduction in Total Factor Productivity (TFP), soil organic carbon depletion, increase in soil salinity, imbalances in water regimes, and pollution due to indiscriminate application of fertilizers and pesticides. Shrinking land resources coupled with accelerated rate of land degradation and ever increasing population driven food demand has drawn attention of the administrators, planners, scientific communities and other stakeholders towards the need of restoring the existing degraded soils so as to utilize the same for increasing productivity. Land degradation is an important global issue because of its adverse impact agronomic productivity, the environment, and its effect on food security and the quality of life (Eswaran, H., Lal, R. and Reich, P. F. (2001). Land degradation: an overview. In: Bridges, E. M., Hannam, I. D., Oldeman, L. R., Pening de Vries, F. W. T., Scherr, S. J., Sompatpanit, S. (eds.). Responses to Land Degradation. Proc. 2nd. International Conference on Land Degradation and Desertification, Khon Kaen, Thailand. Oxford Press, New Delhi, India). Increasingly less remunerative subsistence farming systems lead the cultivators to leave the land uncultivated, which then tends to become barren, and ultimately result in shifting the land resources from agriculture to other uses including industry, which bring environmental degradation. All of these factors combined with increased rate of land degradation are contributing towards decline in agricultural productivity leading to food insecurity. For sustainable development, available land resources need to be utilized based on their potential and limitations. It implies that the productivity of existing degraded soils needs to be restored and that of normal cultivated lands are to be improved. In the context of productivity, land degradation results from a mismatch between land quality and land use (Beinroth, F. H., Eswaran, H., Reich, P. F., Van Den Berg, E. (1994). Land Stresses in Agroecosystems. In: Stressed Ecosystem and Sustainable Agriculture. Virmani, S. M., Katyal, J. C., Eswaran, H., Abrol, I. P., eds. New Delhi: Oxford and IBH).

Under high population pressure, extreme dependence on agriculture, declining soil fertility, climate change and environmental degradation have resulted in backwardness and poverty. Efforts are on in 2013–2014 to increase agricultural production by improving resource efficiency, input use efficiency, ensuring timely availability of quality inputs. The objective is to develop and popularize appropriate ecofriendly farming systems to improve soil health and farm income. The urgent need for conserving our soil and water resources is felt more at present than ever before, because of the fast degrading natural resource base and stagnating crop yield levels. The increase in productivity to meet the demands of growing population and to ensure delivery of food grains as per the terms of proposed Food Security Bill, India needs an additional production of 70 million tons per year. This is possible only by targeting the potential areas and overcoming the constraints in marginal lands. According to the estimates available from various agencies, about 120 m.ha is affected by land degradation, 55 m.ha area lying as Wastelands, 10% of the irrigated lands affected by salinity and alkalinity, widespread deficiency of secondary and micronutrients, over exploitation of ground water, increasing area under current fallows, diversion of prime lands to non-agricultural uses and declining factor productivity threaten the food security of the country.

It is obvious that the neglect and deterioration of the land resources and consequent decline in the productivity is continuing without any check for many years in the country. Continuation of present management practices can damage the soil health very rapidly. If we assume that the situation is more are less similar to Chikkarasinakere hobli of Mandya district, Karnataka, in the entire command area (2.8 Lakh ha) of Cauvery, as the terrain and topography are repeating in the region, then the value of crop loss is about Rs 630 crores and when we add loss due to increased cost of cultivation, the total loss per year is about 795 crores. This excludes the losses to investment/interest on investment made in the creation of irrigation infrastructure by the government. If the interest on investments on irrigation infrastructure is also taken into account, nearly about Rs 1000 crores worth economic loss is taking place due to the development of salinity/sodicity, water logging and multiple nutrient deficiencies in the command area. In the coming years, the situation is likely to worsen further and the expected loss is likely to be higher than what is estimated at present in the command area.

We have to evolve management practices for farmers' choice of remunerative crops without degradation of the natural resource base and define agro-ecological zones where such cropping systems can be adopted in a sustainable manner. India has world's largest livestock population accounting for about 55 and 16% of the world's buffalo and cattle populations, respectively. Rainfed areas account for two-thirds of total livestock population. India ranks second in goats (124.5 m out of world's 764.5 m) while third in sheep (59 m out of world's 1028.6 m), but of late there is a significant change in the livestock composition. India is globally the largest producer of milk with an annual production of 127.9 million tonnes during 2011–2012. Most of this production is coming from the rainfed regions where livestock population is large and is a major component of the livelihood systems.

Currently the deficit in demand and supply of dry and green fodder stands at 12% and 63%, respectively and the gap is likely to be further widened with increase in demand for milk and meat products and change in cropping pattern like cotton and soyabean at the cost of coarse cereals. There is a need to devote at least 10% of the net sown area for fodder production, which is presently around 5% (GoI, 2006).

1.1.2   STATE-OF-PRACTICE OF LAND USE

In Asia, poverty has mainly been rural phenomenon and nearly three-fourth of the poor live in rural areas, with large majority of them dependent on natural resources for employment and income. South Asia, which had a poverty incidence of 43%, contributed about 40% of the world's poor. Development of natural resources thus offers a potentially enormous means of poverty reduction. Nowhere does this situation manifest itself explicitly as in the states of Bihar, Uttar Pradesh, etc.

Land utilization is a complicated issue and to some extent represents the man-environment relationship. In addition to the role of land and water, in terms of quantity as well as quality, the other important aspects in any land utilization are the management practices, the human dimension which influences and ultimately decides the success or efficiency of land use. Any development is sustainable only when an optimal and efficient plan to utilize natural resources is firmly in place. In this context it is widely perceived that any State can only prosper and have sustainable growth pattern if it utilizes its natural resources properly and judiciously. Otherwise even with the most productive soils and abundant water, economic poverty and backwardness will prevail. In the backdrop of critical issues, that is, climate change, food security and self-reliance, use and reliability of information technology tools, available and proposed legal framework, mobilization, capacity building and knowledge enhancement related to land use planning it is a challenge and uphill task for policy makers and planners to recommend an acceptable and implementable land use policy framework.

Any development is sustainable only when an optimal and efficient plan to utilize natural resources is firmly in place. Today the research focus is on the dynamics of land use. Land represents an important resource for the economic life of a majority of people in the world. The way people handle and use land resource is decisive for their social and economic well-being as well as for the sustained quality of land resources. This implies that policy discussion and development planning need to be based on a sound understanding of these dynamics. Conflict over land use is inevitable. With high population growth, endemic poverty and weak existing institutional capacity for land management a scientific and strategically important land use policy is must for sustainable development.

Planning and management of land resources are integral part of any developmental program. The importance of land as a resource cannot be overemphasized and one of the major global concerns is the problem of declining agricultural pro-

ductivity and land resources that are being threatened by the rapid human population growth. Sustainable land and natural resource management is a key factor in
ensuring adequate food, fodder, and fiber production. Land use intensification due
to demographic pressure led to expansion of agricultural land in to fragile systems
and it is one of the major causes of degradation of natural resources. Considering
the multiple demands of land and water resources, the time has now come to take
hard decisions in planning limited natural resources right from the lowest possible
level to fulfill the aspiration of all the stakeholders. It provides a scientific basis
for resource management, conflict resolution and decision-making at the local and
regional levels, consistent with national policies and priorities. However, the existing database on land use is highly inadequate. Therefore, strengthening of the
database, using traditional cadastral surveys, modern remote sensing techniques,
GPS, GIS and computerization of land records would be necessary. In this context,
identification and delineation of prime agricultural lands helps land use managers to
take appropriate decisions for sustainable agricultural production and food security
besides conserving precious natural resources. Conversion of agricultural land to
non-agricultural uses is more intense in recent years in India. The causes for loss of
agricultural lands are almost self-evident. They include, profits from land sales for
non-farming purposes, conversion of agricultural lands for housing, and development of highways, industries, construction of airports, etc. Prime agricultural lands
need to be preserved for future food security and prosperity of rural areas. Delineation of prime agricultural lands at different levels helps in spatial zoning policy
(Special Agricultural Zones), where zoning of land use ensures and stimulates optimal use of the land on a permanent basis. Such preservation strategies require a
vision of how agriculture should develop and which farmland should be preserved
in order to achieve this vision. In general, prime agricultural lands have an adequate
and dependable water supply from precipitation or irrigation, a favorable temperature and growing season, acceptable acidity or alkalinity, acceptable salt and sodium
content, and few or no rocks. They are permeable to water and air. Prime farmlands
are not excessively erodible or saturated with water for a long period of time, and
they either do not flood frequently or are protected from flooding (Soil Survey Staff,
1993).

1.1.3 MAPPING THE RESOURCES

In late sixties it was proposed to classify into 28 broad soil classes and suggested
that composite map of each village be prepared in the scale 1: 7920 or 1: 3960 giving boundaries of land use classes and cropping pattern (Raychaudhuri, S. P., 1968,
Interpretative Soil Groupings – Prediction of Soil Behavior and Practical use of Soil
Maps, J. Indian Soc. Soil Science P. 205–213; Raychaudhuri, S. P., Govindarajan, S.
V. (1971) Soils of India. ICAR Tech. Bull. (AGR) No. 25, 45 pp). This can be obtained only by conducting cadastral level detailed surveys wherein we characterize

and delineate homogenous areas based on soil-site characteristics into management units. During the period 1986–1999, the National Bureau of Soil Survey and Land Use Planning (NBSS & LUP) of the ICAR completed a Nation-wide soil survey at 10 km grid interval and brought out soil map of different states in 1:250,000 scale. Detailed soil survey available at 1:10,000 scale using cadastral map as the base will lead to the preparation of village level/watershed level soil maps with land management units delineated for adoption of appropriate soil and crop management practices. A case study based on this approach for the soils of Andhra Pradesh has been brought out (Naidu, L. G. K. et al., 2013, Need for Management Domain Soil Maps: A Case study of Andhra Pradesh. Indian Journal of Agropedology (in press). The Indian Society of Soil Survey and Land Use Planning). This paradigm shift from soil mapping to land resource mapping at village level by detailed soil survey and socio-economic survey is suggested for embarking upon a National Mission Mode Project, which will help in the development of soil productivity indices and rating, land capability classification and extent of degraded lands at village level and development of a National Portal on soils of India for knowledge based Agro-technology transfer by modern methods of ICT for ensuring sustainable agriculture and food security for the people now and in the future (Velayutham, M., 2012, National Soil Information System (NASIS) and Land Resource Mapping for Perspective Land Use Planning and Pragmatic Farm Level Planning, Madras Agric. J., 99(4–6), 147–154).

The use of modern geospatial technologies such as high resolution satellite data, GPS and GIS can be effectively used in land resources inventory, mapping, monitoring and management. The satellite data provides integrated information on landforms, geological structures, soil types, erosion, land use/land cover, surface water bodies and qualitative assessment of groundwater potentials at watershed level. The applications of space technology are rapidly advancing in land resources inventory, mapping and generation of databases on a regular basis for better planning, management, monitoring and implementing the land use plans at large scales. Satellite remote sensing data from LANDSAT-ETM+, IRS-IC, IRS-ID, IRS-P6, Cartosat-I, Cartosat-II, Quickbird and Google are available for generation of spatial database on land resources for various applications. GIS techniques are playing an increasing role in facilitating integration of multi-layer spatial information with statistical attribute data to arrive at alternate developmental scenarios in sustainable management of land resources. The indispensable role of satellite remote sensing in agriculture sector was fully appreciated ever since the launch of Landsat 1, the first civilian remote sensing satellite in July 1972. In fact, the choice of spectral bands in most of the civilian remote sensing satellites launched by various countries, including India, to-date has been dictated by their use for efficacious management of resources related to agriculture. This capability has been further enhanced manifold by Geomatics during the last few years, especially with the availability of high-resolution satellite images from IRS and Cartosat satellites. Geomatics deals with the tools/techniques related to the measurement of geographical data, the data that has both spatial and non-spatial components. Most common among these tools are remote sensing, geographical information system (GIS) and Global Positioning System (GPS). Hence, in the twenty-first century, to steer the agricultural achievements towards the path of an 'evergreen revolution,' there is a need to blend the traditional knowledge with frontier technologies. Information and communication technology; space technology; geographical information systems (GIS) are the tools of such frontier technologies which would help in creating agricultural management systems; making plans for sustainable agriculture; and bringing new areas (through development of wastelands) into productive agriculture.

1.2 NEED OF THE HOUR

Lessons learnt in the form of both success stories and failures may be shared with the local extension systems and development departments for ensuring continuity of follow up action. Systematic study of databases on natural resources should aid formulation of coherent policies for optimal use of the resources in an eco-friendly manner. A perspective land use policy and location specific land use plan with SWOT analysis may be evolved. Information needs on Natural resources for development of Land Use Plan (LUP) is a pressing need of a "National Portal on Natural Resources" of digitized land resource maps of not less than 1:10,000 scales for realistic land use at farm watershed/village/District/State levels.

2 OBJECTIVE

To suggest strategies and programs for optimizing agriculture land use and rural development.

2.1 STATUS

Watershed based Planning and Synergies in Land Use: Landscape Plan for optimizing agriculture land use and rural development are described.

2.1.1 WATERSHED BASED PLANNING

As India is one of the developing countries, due to increased urbanization and deforestation the vegetative cover is being decreased as a result of which the hydrological cycle is being affected and this is resulting in bottoming levels of the ground water. All these consequences finally result in desertification of that piece of land. Watershed planning is very much essential for the conservation of land and water resources and their management for optimum productivity. Reliable and timely information on the available natural resources is very much essential to formulate a comprehensive land use plan for sustainable development. The land, water, minerals and biomass resources are currently under tremendous pressure in the context of highly competing and often conflicting demands of an ever-expanding population. Consequently over exploitation and mismanagement of resources are exerting detrimental impact on environment.

A broader, integrative or holistic view, takes into account both, a vertical aspect – from atmospheric climate down to ground water resources, and a horizontal aspect an identifiable repetitive sequence of soil, terrain, hydrological, vegetative and land use elements. Soil, water, flora and fauna are the important land resources, which together influence the survival of human beings by supporting food production and providing a congenial living environment. There was harmony between the mankind, the other living beings and the surrounding environment. Increasing needs of the growing population inevitably lead to agriculture and Industrial Development through haphazard usage of land resources, which eventually disturbed the ecological balance through pollution and declining quality of soil, water and atmosphere. In developing countries, and especially in the most densely populated countries like India, there is an urgent need for immediate action as there will be pressure to make ad-hoc decisions and to push forward development schemes, for social or political reasons.

A systematic assessment of land and water potential, alternatives for land use, and economic and social conditions is necessary for selection and adoption of the best land-use options. Hence, it is imperative to assess the development potential of the dominant sector of the economy, that is, agriculture. Several attempts made to study the agrarian structure and economic development has failed to identify the prime reasons behind the dismal performance of the farm sector especially at the micro level. Attempts to assess the suitability of micro-climatic and region specific socio-economic parameters in farming are often staggered and noncomprehensive. Dovetailing soil survey data with other natural resources data and socio-economic parameters, macro-level planning of land use for different sectors can be prepared at levels of Agro-ecological regions, Agro-ecological sub-regions and Agro-ecological zones at state level for development planning. Soil survey information with other natural resources and socio-economic factors needs to be incorporated in GIS framework for developing integrated land use and management predictions and validation. For rational utilization of available natural resources and for taking up timely and reliable information on natural resources with respect their nature, extent and spatial distribution; and nature, magnitude is required.

In the words of eminent economist, Professor C.H. Hanumantha Rao, "Watershed development has been conceived basically as a strategy for protecting the livelihoods of the people inhabiting the fragile ecosystems experiencing soil erosion and moisture stress." This strategy has been designed with the objectives of public participation for conservation, upgradation and utilization of natural endowments (i.e., land, water, animal and human resources) in a harmonious and integrated manner, generation of massive employment in rural areas, improved standard of living of millions of poor farmers and landless laborers, restoration of ecological balance through scientific management of land and water, reduction of inequality between irrigated and Rainfed areas. The Ministry of Agriculture has established the National Rainfed Area Authority (NRAA) to meet this objective. The Planning Commission has published a Common Guidelines for undertaking Watershed Development Program (http://www.nraa.gov.in). Initially, this NRAA was established under the Ministry of Agriculture, and then shifted to the Ministry of Rural development, and now under the Planning Commission.

Since 1994, the country has wide variety of experiences in facilitating the watershed program. The Nagpur Declaration (Indian Geographers Congress, organized by the National Associations of Geographers India (NAGI), held at NBSS&LUP (Nagpur), during 1–3, January 2000) on "Natural Resources Planning and Management for Sustainable Development" suggested that both "river basins management" at the macro level and "watershed management" at the micro level should mutually complement each other, for integrated Water resources planning and management. In India in the last three decades, watersheds have become the pivotal unit for rural development programs. Growing concern of poverty, population growth and environmental degradation have led to increasing public investment in India towards integrating resources through watershed management. The term 'watershed development' has ideally been accepted "as a geophysical unit for planning and executing development program for rational utilization of all natural resources for sustained optimum production of biomass with the least damage to the environment" (GoI, 1999:ix). However, due to high importance placed to about 53% of the total geographical area subjected to degradation, soil and water conservation measures have been adopted towards improving agriculture production and productivity to cater to large percentage (70%) of rural population. Eventually, the Guidelines of watershed development program are revised in 2001 (Watershed Guidelines – Revised) and 2003 (Hariyali). Apart from these guidelines, Ministry of Agriculture also issued guidelines for National Watershed Development Project for Rainfed Areas (2000). The potential of Watershed becoming a functional unit remains in understanding the linkages and promoting economic activities that sustain watershed management. A participatory watershed management approach is considered as the ideal for achieving food security and sustainability (Yoganand, B., Tesfa Gebremedhin, 2006, — Participatory Watershed Management for sustainable rural livelihoods in India, Research paper; Vidula, A. S., Kulkarni, S. S., Santosh Kumbhar, Vishal Kumbhar, 2012). Participatory Watershed Management in South Asia: A Comparative Evaluation with Special References to India, International Journal of Scientific and Engineering Research, Vol. 3(3). WASSAN (2003) Hariyali – Issues and Concerns).

The most disturbing fact is that almost all the estimates are based on sample survey only. At present the Govt. of India has not made it obligatory for the generation of farm wise detailed land resources database for the agencies executing the watershed development programs. There is a need to give legal status/backup by Government of India, so that all the land based rural development programs are based on the farm specific land resources database and are truly science based.

2.1.2. SYNERGIES IN LAND USE—LANDSCAPE PLAN

In a land scarce country like India agriculture can no longer be treated as an isolated, single function land use. Any guidelines (Dent, D., Young, A., 1981. Soil Survey and Land Evaluation. London, Allen and Unwin. 278pp.) cannot be a prescription for land use planning, it should be flexible involving local or national procedures that can be modified by the people on the spot. A paradigm shift is necessary to overcome the impending challenges of continued competing demand for land for different uses without sacrificing other goals such as environment services, recreation, etc. Landscape approach is possibly the best way to deal with the natural resources based rural development. A coalition of agricultural and environmental organizations has released a new report recommending an "integrated or whole landscape" approach. It's an approach that brings together diverse and competing groups to discuss common needs. "This whole approach came out of the realities of competing interests – that in agricultural landscapes it's not just the farmers that want the water and it's not just one set of farmers that want the water. That maybe the crop farmers are competing with the livestock farmers for water and they're competing with the local towns for water. So a lot of these alliances that we're talking about started off as conflict situations or at least a lot of tempers being raised on a regular basis," (Sara Scherr, an agricultural and natural resource economist and founder, president and CEO of Eco-Agriculture Partner "http://www.voanews.com/content/decapua-africa-landscapes-19jun12/1212884.html"). Even landscape planning and land use zoning approach can co-exist depending on the acceptability among the stakeholders.

The land resource evaluation forms a pre-requisite for sustainable land use planning. Land use planning aims to encourage and assist land users in selecting options that increase their productivity, are sustainable and meet the needs of society (FAO, 1993. Guidelines for Land Use Planning. FAO Development Series 1, FAO/AGLS, Rome, 96 p). Later urban developers found the mixed – or multi-use developments appealing because such developments offered a way to capture several types of development in one project that was larger than any single project they might create in the same place. Moreover, the interaction and sharing of facilities had the potential to reduce long-term development costs and increase profitability. Development of urban agriculture in west is one of such development. The key to success is synergy between the land uses.

The key is the effect of synergy must result in multiplier effect of combined/mixed land use. For instance, benefits of mixed use of land for agriculture and forest should be more than the sum of individual benefit. A mixed and integrated rather than a sectoral approach is an effective way of preventing or resolving the conflicting land uses. It optimizes planning process by involving all kinds of stakeholders as well. Thus rural development in India account for either agriculture or forest disregarding the fact that these two land uses co-exist in almost half the geographical area of the country and play significant role in livelihood of the inhabitants.

Some states (Tamilnadu and Chhattisgarh) started offering subsidy for energy plantation on marginal lands. Land use policy has been shown to be important in shaping bioenergy production (Rokityanskiya, Dmitry, Pablo C. Benítezb, Florian Kraxnera, Ian McCalluma, Michael Obersteinera, Ewald Rametsteinera, Yoshiki Yamagatac, 2007. Geographically explicit global modeling of land-use change, carbon sequestration, and biomass supply, Technological Forecasting and Social Change, 74(7), 1057–1082. Gillingham, K. T., Smith, S. J., Sands, R. D., 2008. Impact of bioenergy crops in a carbon dioxide constrained world: an application of the MiniCAM energy-agriculture and landuse model. Mitigation and Adaptation Strategies for Global Change 13, pp. 675–701; Wise, M. A., Calvin, K. V., Thomson, A. M., Clarke, L. E., Bond-Lamberty, B., Sands, R. D., Smith, S. J., Janetos, A. J., Edmonds, J. A., 2009. The implications of limiting CO_2 concentrations for land use and energy. Science 324, 1183; Melillo, J. M., Reilly, J. M., Kicklighter, D. W., Gurgel, A. C., Cronin, T. W., Paltsev, Felzer, B. S., Wang, X., Sokolov, A. P., Schlosser, C. A., 2009. Indirect Emissions from Biofuels: How Important? Science, 326(5958), 1397–1399).

The basis of this (landscape) planning approach (Hawkins, V., Selman P., 2002. Landscape scale planning: exploring alternative land use scenarios Landscape and Urban Planning Volume 60, Issue 4, 30 August 2002, Pages 211–224) is to map at various scales those elements in the landscape that are inherently stable or unstable, and to determine from these maps a network of landscape elements to act as 'biocentres' and 'biocorridors' (Bucek, A., Lacina, J., Michale, I., 1996. An Ecological Network in the Czech Republic. Veronica special 11th issue. Dent, D., Young, A., 1981. Soil survey and land evaluation. London, Allen and Unwin. 278 pp.). The existing network can then be analyzed to identify where landscape creation or rehabilitation is necessary to fill strategic gaps. The basic concept involves retaining existing ecological infrastructure, and then creating 'more of the same' landscape elements in deficient areas.

2.2 NEED OF THE HOUR

Developing a "Greening Economy" where farm/social forestry can be developed so that requirements of forest dependent populations industry as well as consumers can be taken care off. The benefits should continue to accrue even after the completion of the program and the structures are maintained on a continuing basis. Contingency plans should be integral in land use plans. Associated support services in the form of community seed banks may be strengthened. Agricultural marketing services may be expanded to diverse production zones. Social auditing of all the Agriculture and Rural Development programs may be undertaken at watershed/district level. Information Service on "Integrated Land Use Planning for Sustainable Agricultural," is to be introduced. Forward and backward linkages may be strengthened between Special Zones and Peri-urban areas to ensure the Demand-Supply Chain.

Promote participatory management and efficient conservation and utilization of the Natural Resources for Sustainable Agriculture. Joint Action/ Operational Research and field demonstration programs may be strengthened. Train youth, progressive farmers, Panchayat representatives, women and others through "Quality training materials and case studies." Ministry of Environment and Forest Guidelines, Ecosystem-specific on-farm and non-farm activities and land management programs, needs awareness, education, field visit, training and demonstration programs. Transparency in decision making by display of Natural Resources and cropping pattern at states, districts, blocks, villages and Panchayat Offices.

3 OBJECTIVE

To identify policy related issues for optimizing agriculture land use contributing to rural development.

3.1 STATUS

Described here under are Policy Supports, National Land Use Policy and Crop Insurance.

3.1.1 POLICY SUPPORTS

The National Commission on Agriculture (1972) and the task forces constitute by the Planning Commission in 1972 and 1984 have emphasized the need for systematic and detailed soil survey of the agricultural lands at large scale mapping for scientific crop and land use planning. Land and Water are subjects within the purview of the States. However, 'Forest' was a State subject earlier, and was brought to the Concurrent list in 1976. The subject 'Environment' is not under any List but is covered under the Directive Principles of State Policy and Fundamental Duties enshrined in the Constitution 'to protect and improve the environment.' The list of central policies which have a bearing on land use, include National Water Policy, 1987; National Land Use Policy Outlines, 1988; National Forest Policy (NFP) of 1988; Policy Statement of Abatement of Pollution, 1992; National Livestock Policy Perspective, 1996; National Agricultural Policy, 2000; National Population Policy, 2000; National Policy and Macro-level Strategy and Action Plan on Biodiversity, 2000; and National Land Reforms Policy, 2013. In addition, there are legislative frameworks, which have to be conformed to by any state while planning a land use policy. These include: Forest (Conservation) Act, 1980; Environment (Protection) Act, 1986; Water (Prevention and Control of Pollution) Act, 1974 as amended in 1988; Wildlife (Protection), 1972 as amended in 1988; Constitutional Amendments (73rd and 74th Amendments) of 1992; and Municipality Act, 1992 (74th Amendment Act, 1992).

In order to further revive agricultural growth in the states, agro-ecological planning/agro-climatic planning could be adopted in a decentralized planning framework to optimally utilize the zonal potential/resources in crop and livestock production." (Planning Commission, GOI, 2011). Non-availability of accurate and dependable data base below the taluk level on agriculture and allied sector has been pin-pointed as one of the major constraints impeding resource-based planning by the Planning Commission, GOI (2011). They have rightly pointed out the need for initiating comprehensive exercise at state level to delineate the state into various agro-ecological/agro-climate zones and the districts into further sub-units. In each agro-ecological unit, resource based plan need to be developed including issues like yield gap also. The Agro-climatic Regional Planning exercise of the Planning Commission during 8th and 9th plan period was a serious attempt to plan for smaller homogenous regions, keeping in view the natural resources and capabilities towards achieving development. The Agenda 21 Document of the Rio Conference (1992) upholds the basic right of people to be involved in decision-making exercises, which directly affect them. The problems and constraints to farming were assessed along with the strength and opportunities to develop farming into a sustainable activity on the basis of resource endowments in the respective agro-ecological units. Enterprise mixes were suggested for the agro-ecological units based on stakeholder preferences and suitability.

The Mission Document of National Mission on Sustainable Agriculture (NMSA) 2009, and 2013, one of eight missions are: devise strategies to make Indian agriculture more resilient to climate change, support the convergence and integration of traditional knowledge and practices systems, update and inventories of available region-wise technologies for farming system, generate awareness through stakeholder consultations, training workshops and demonstration exercises for farming communities, for agro-climatic information sharing and dissemination, strengthen the National Seed Grid ensuring supply of seed across the country as per area specific requirements, issue Soil Health Pass Book to each farmer with soil testing advisories within next five years, and undertake, in an organized manner, the production and marketing needs of Bio-fertilizers, liquid fertilizers and compost, etc. Under this Sub-Mission, the Department of Agriculture and Cooperation (DAC) proposed to collect data on agro climatic variables, land use and socio economic factors through NIC who in turn will use the database so developed as well the data warehouse developed by IASRI and other relevant sources such as IMD and National Centre for Medium Range weather forecasting (NCMRWF), etc., of the

Ministry of Earth Sciences for developing user friendly software tools and necessary hardware, integration into web based framework and development of decision support systems (DSS).

The National Policy for Farmers (2007) provides for preparation of Drought Code, Flood Code and Good weather Code for drought prone, flood prone and other agricultural areas respectively, so that the required corrective/mitigation measures can be taken well in time. The National Action Plan on Climate Change (NAPCC), has dealt with "Information and Data Management" issue, in great details and also recommended database development projects for operationalization during 2012–2017 by creation of detailed Soil Data base to develop micro level agricultural land use plan and Space enabled spatial database for Village Resource Centers and Development of GIS and remote-sensing methodologies for detailed soil resource mapping and land use planning at the level of a watershed or a river basin.

The legislations like 'The Scheduled Tribe' and other Traditional Forest Dwellers' (Recognition of Forest Rights Act) 2006, the National Policy for Farmers 2007 and other new anthropogenic policy initiatives have also necessitated a debate on the need for synergies between forest and agriculture. While the former has institutionalized and legalized agriculture practices in forest areas the latter has now included Pastoralists/Herders/Migratory Glaziers/Tribal Families/Persons engaged in shifting cultivation and in the collection, use and sale of minor and non-timber forest produce also in the list of farmers.

The amendments of the constitution transferring the subject of land from state list to the concurrent list needs due consideration (Komawar and Deshpande, S. B., 2008. Legal Aspects of Land Use Planning and Policy. National Brain Storming Session on Land Use Planning and Policy Issues at NBSS&LUP (ICAR), 25–26 July 2008, Nagpur).

3.1.2 NATIONAL LAND USE POLICY

There is an estimate of declining agriculture landholding to 0.68 ha by 2020, and to a low of 0.32 ha in 2030. Out of 140 million hectares is net cultivated area, about 57 million ha (40%) is irrigated and the remaining 85 million ha. (60%) is rainfed. This area is generally subject to wind and water erosion and is in different stages of degradation due to intensive agricultural production. Therefore, it needs improvement in terms of its productivity per unit of land and per unit of water for optimum production.

A key element of the Farmer Centered Agricultural Resource Management (FARM) Program, an initiative of eight Asian countries including China in 1990's supported by UNDP and implemented by the FAO is the recognition of communities "indigenous knowledge, which when complemented by specialist "formal knowledge," promote participatory learning and research for achieving sustainable use and management of natural resources in agriculture and attainment of household food security through innovative approaches.

The "National Land Use Policy guideline and action points" were prepared by the Government of India, Ministry of Agriculture after intensive deliberations In the said policy, framing of suitable legislation and its sincere enforcement were stressed by imposing penalties, of violation thereof. The said policy guidelines were placed before the 'National Land Use and Waste Land Development Council' under the chairmanship of Prime Minister and its first meeting was held on 6th February, 1986. The Council agreed to the adoption of policy and circulated the same throughout the country for adoption after suitable considerations at the state level. Of the nineteen points, some of the most important ones are: Land Use Boards at the State level should be revitalized; Land Use Policy must be evolved by all users of land within Government jointly and must be enforced on the basis of both legislation for enforcing land use as well as their promotional and preserving methods; Urban Policy must be restructured so as to ensure that highly productive land is not taken away. Town planning should also provide for green belts; A national campaign should be launched for educating the farmers and Government Departments about the need to conform to an integrated land use policy; Land and soil surveys should be completed and inventory of land resources should be prepared in each State so that resources allocation is based on a reliable data base; Heavy penalties should be imposed against those who interfere with land resources and its productivity. It must be recognized that environmental protection cannot succeed unless this is done; The problems of water logging, salinity and alkalinity must be brought under control by the use of appropriate technologies and by the adoption of proper water management practices; Land Use Planning should be integrated with rural employment programs in such a manner that loans and subsidies are given only for those productive activities which represent efficient land use; Rights of tribal and poorer sections on common land should be protected through legal and administrative structures, etc. The policy was finally adopted in 1988 but has not really been able to move things. The Policy has been circulated to all concerned for adoption and implementation through enactment of suitable legislation. The policy, however, did not make the desired impact, mainly due to the fragmented handling of different components of agriculture like land and soil. The framework for the Land Use Policy to be developed by each state is already available in the Guidelines for Land use Policy (1988) of the Government of India. The states are now required to fit in the particular scenario that exists in the region.

Any policy or regulating act prepared for a region should conform to the prevalent situation of the area. The basic purpose of any such piece of legislation is to protect the natural resources from being over-exploited at any given time and also to improve the living conditions of the inhabitants. Land related policies and regulating acts, however, are perceived differently as they restrict and regulate activities, which may be beneficial to individuals at the moment, but are detrimental to the cause of society in the long run. It is here that the long term benefits of sustainability of resources needs to be emphasized. The market driven land use pattern may yield higher returns in the short run, but may pose several unmanageable problems for future generations due to unplanned overexploitation of land, water and other natural resources. Land use decisions are not made just on the basis of land suitability but also according to the demand for products and the extent to which the use of a particular area is critical for a particular purpose. On several occasions only 50% of the farmers adopted land use that is considered scientifically correct or recommended by the NARS.

The issues are diverse and challenging in other agro-ecosystems viz, arid, irrigated, coastal and hill and mountain regions. Rainfed areas show poor socio-economic status such as limited irrigation facility, 15% as compared to 48% in irrigated regions; lower employment opportunities and higher population of agricultural labor force, lower productivity and lower per capita consumption, poverty, seasonal out-migration, poor infrastructure and social developmental indices. Apart from above, population pressure, fragmentation of land holdings, tenancy farming, and low investment capacity, low productivity of crops and income, credit, pricing policy, marketing and wage rates are the major socio-economic issues that impact rainfed agriculture (Ravindra Chary, G., Vittal, K. P. R., Ramakrishna, Y. S., Sankar, G. R. M., Arunachalam, M., Srijaya, T., Udaya Bhanu, 2005. Delineation of Soil Conservation Units (SCUs), Soil Quality Units (SQUs) and Land Management Units (LMUs) for Land Resource Appraisal and Management in Rainfed Agroecosystem of India—A Conceptual Approach. Lead Paper. Proceedings of National Seminar on Land Resources Appraisal and Management for Food Security, organized by Ind. Soc. of Soil Survey and Land Use Planning, NBSSLUP, Nagpur, 10-11, April).

The issues regarding SEZs vis-a-vis agricultural land use and rural development are of major concern today and need to be discussed in a food, nutritional and rural livelihood security context. Though significant proportion of the rural population is expected to move to cities by 2030, still substantial numbers will remain in rural areas. It is envisaged that leasing of the land and the pooling of the holdings may become more common in future.

3.1.3   CROP INSURANCE

Agriculture is under tremendous pressure due to ill effects of climate change. Agriculture and forestry sectors contribute around 1/3rd of global warming potential and are also highly sensitive to climate change. The impending climate change/variability is likely to accentuate the production related problems both in, irrigated, rainfed and other regions. Both rainfed agriculture and rural development are land-based activities, hence an integrated land use planning therefore becomes necessary to achieve sustainable rainfed agriculture and rural development. Developing country like India is particularly vulnerable because its population depends directly on agriculture and natural eco systems for their livelihoods. The warming trend in India over the past 100 years (1901 to 2007) was observed to be 0.51°C with accelerated warming of 0.21°C per every 10 years since 1970 (Krishna Kumar. 2009. Impact of climate change on India's monsoon climate and development of high-resolution climate change scenarios for India. Presented at MoEF, New Delhi on October 14, 2009. http: moef.nic.in). A more recent study indicated that annual minimum temperature is increasing @ 0.24°C per 10 year on all India basis (Bapuji Rao, B., Santhibhushan Chowdary, P., Rao, V. U. M., Venkateswarlu, B., 2013. Rising minimum temperature trends over India in recent decades: Implications for agricultural production. Current Science, Communicated). The projected impacts are likely to further aggravate yield fluctuations of many crops with impact on food security and prices. Climate change impacts are likely to vary in different parts of the country. Parts of Western Rajasthan, Southern Gujarat, Madhya Pradesh, Maharashtra, Northern Karnataka, Northern Andhra Pradesh, and Southern Bihar are likely to be more vulnerable in terms of extreme events (Mall, R. K., Gupta, A., Singh, R., Singh, R. S., Rathore, L. S. (2006a). Water resources and climate change: An Indian perspective. Current Science. 90(12), 1610–1626; Mall, R., Singh, R., Gupta, A., Srinivasan, G., Rathore, L. (2006b). Impact of Climate Change on Indian Agriculture: A Review. Climatic Change. 78, 445–478). Hence, there is a need to address the whole issue of climate change and its impacts on Indian agriculture in totality so as to mitigate the same through adaptive techniques against the global warming on war-footing.

The idea of crop insurance emerged in India during the early part of the twentieth century. Yet it was not operated in a big way till recent years. J.S. Chakravarti proposed a rain insurance scheme for the Mysore State and for India as a whole with view to insuring farmers against drought during 1920s. Crop insurance received more attention after India's independence in 1947. The subject as discussed in 1947 by the Central Legislature and the then Minister of Food and Agriculture, Dr. Rajendra Prasad gave an assurance that the government would examine the possibility of crop and cattle insurance. In October 1965, the Government of India decided to introduce a Crop Insurance Bill and a Model Scheme of Crop Insurance in order to enable the States to introduce, if they so desire, crop insurance. In 1970, the draft Bill and the Model Scheme were referred to an Expert Committee headed by Dr. Dharm Narain. Different experiments on crop insurance on a limited, ad-hoc and scattered scale started from 1972–1973.

Agriculture Insurance, including livestock insurance has been practiced in the country for over 25 years. Prior to 2002–2003 General Insurance Corporation of India (GIC) was implementing National Agricultural Insurance Scheme (NAIS). Recognizing the necessity for a focused development of crop insurance program in the country and an exclusive organization to carry it forward, Government created an exclusive organization—Agriculture Insurance Company of India Limited (AIC) on 20th December 2002 (www.aicofindia.org). AIC commenced business from 1st

April 2003. AIC introduced rainfall insurance known as 'Varsha Bima' during the 2004 South-West Monsoon period. Varsha Bima (Varsha Bima covers anticipated shortfall in crop yield on account of deficit rainfall. An analyzis of Indian Crop Insurance Program between 1985 and 2003 reveals that rainfall accounted for nearly 95% claims – 85% because of deficit rainfall and 10% because of excess rainfall (AIC. 2006. Crop Insurance in India. Agriculture Insurance Company of India Limited (AIC), New Delhi, p.10).

3.2   NEED OF THE HOUR

Land use has become a national concern in the Globalization era and is being impinged upon by International Trade Agreements. Therefore, there is a need to amend the Constitution transferring the subject of land from State list to Concurrent list to meet the targets of production, distribution and trade. Viable land use planning and its implementation needs land reforms in land tenure, rights, title and lease deeds by enacting appropriate legislative measures.

States have to study the implications of an optimal land use plan and evolve policies and strategies to regulate the market distortions, marketing arrangements, processing and price supports. Greening rural economy is integral part of such policies. Weaning away increased use of marginal lands for agricultural production and arresting diversion of good agricultural lands for non-agricultural purposes may form part of land use policy. Ensure livelihoods of forest dwellers while protecting bio-diversity forest based livelihoods like Medicinal and Aromatic Plants, Bee Keeping, Tendu leaf collection, Non Timber Forest Produces (NTFPS). Land use policy should promote copping based on soil fertility – water resources and climate changes. Incentives should be provided for to Reallocate land of small and marginal wherever necessary. In both the underground and open cast mining, it is necessary to develop reclamation and rehabilitation strategies to a productive use as soon as possible after mining. Improvise institutional units and Joint operational mechanisms for efficient management of Natural resources.

Agro-meteorology Advisory services at village level may be strengthened to provide information on weather forecasting and action plans. Weather insurance protocols may be standardized and operationalized at community level.

4   OBJECTIVE

To develop a framework for convergence of agriculture and related rural development programs for sustainable land management.

4.1 STATUS

Hereunder various Government Agencies and Action Mapping are described.

4.1.1 GOVERNMENT AGENCIES

Various schemes and programs were initiated by central and state governments under the aegis of Planning Commission and tried to address either directly or indirectly the LUP in sectoral manner. For example, programs like Multi Cropping Program, Command Area Development Program (CADP), Hill Area Development Program (HADP), Desert Development Program (DDP), Drought Prone Areas Program (DPAP), Whole Village Development Program, Decentralized Village Scheme, Integrated Waste Land Development Program (IWDP), Integrated Watershed Management Program (IWMP), Accelerated Rural Water Supply Program (ARWSP), accelerated Irrigation Benefits Program (AIBP) and Mahatma Gandhi National Rural Employment Guarantee Scheme (MGNREGS) of Ministry of Rural Development, GoI, Backward Region Grant Fund (BRGF) Program of Ministry of Panchayat Raj, GoI, and Comprehensive Land Development Program (CLDP) in Andhra Pradesh, etc. Many of these programs though have land based rural developmental activities but were implemented in a piece meal and top-down manner, many a times without considering the features of the locally existing natural resources (particularly land resources). As agricultural policies and water management strategies evolved over the years, water-use trends changed accordingly.

Government of India spends every year nearly Rs one lakh crore on various agricultural development programs, soil and water conservation being the most prominent activity. However, various post project evaluation reports suggest that 50–58% of farmland areas are suffering from various kinds of degradation. It indicates that there is mismatch between the actual requirement and what is actually implemented in the field in the form of various land based rural development programs. Such situations can be avoided if detailed, farm specific land resources database are generated in each and every project area to identify inherent potentials and constraints and evaluated for suitability to various land uses. By acting as driver of change in land use, the massive irrigation systems have added another dimension to the twin problems of water logging and secondary salinization. Earlier studies have revealed that the small and marginal farmers are worst affected by soil degradation (Jai Singh and Jai Pal Singh 1995. Land degradation and economic sustainability, Ecological Economics. 15A (1), 77–86).

4.1.2 ACTION MAPPING

It has long been felt that our natural soil environment should be mapped and monitored with active participation of agencies responsible for managing natural resources, industry groups and community organizations. It would not only store the datasets for posterity but also will improve our understanding of biophysical processes in terms of cause-effect relationship in the pedo-environment to understand soil health. Soils, which previously had a high nutrient, are now found to be deficient in many nutrients. Long-term fertility studies have shown reduction in organic matter and other essential nutrients (Bhandari, A. L., Ladha, J. K., Pathak, H., Padre, A. T., Dawe, D., Gupta, R. K. (2002). Yield and soil nutrient changes in a long-term rice-wheat rotation in India. Soil Sci. Soc. Am. J. 66, 162–170). This indicates that biological activity of soils in maintaining the inherent soil fertility has been reduced over the years of cultivation. Thus the soils need use of chemical fertilizers in maintaining their fertility (Abrol, I. P., Gupta, R. K., 1998. Indo-Gangetic Plains – Issues of changing land use. LUCC Newsl., March 1998, No. 3).

For rationalized optimal utilization of available natural resources and for taking up scientific decisions on any preventive or curative measures, timely and reliable information on natural resources with respect to their nature, extent and spatial distribution; and nature, magnitude and temporal behavior of various types of degraded lands, and temporal behavior of various land use features is a pre-requisite. Diversity in the micro-environment imposes severe limitations to introduce a plan for the farming systems. A systematic assessment of physical, social and economic factors is necessary to encourage and assist land users in selecting sustainable options that increase their productivity and meet the needs of society. Specific agro-ecological unit based interventions with respect to crop, livestock production and other agro-related enterprises and spatial integration of crop is warranted. Integration of other line departments and major programs like NREGS under the aegis of Krishibhavans with the support of local bodies is suggested.

The Department of Agriculture and Cooperation (DAC) has taken steps to digitize "the survey data as well as the relevant maps" by the Soil and Land Use Survey of India (SLUSI), in collaboration with National Informatics Centre (NIC). The National Informatics Centre (NIC) has already published its village level dataset designed, for its Projects for grassroots development: (a) DISNIC-PLAN Project: IT for Micro Level Planning (http://disnic.gov.in) and (b) Agricultural Resources Information System (AgRIS) Project (http://agris.nic.in) of Department of Agriculture and Cooperation pertains to Soil, Land, Groundwater and Environment parameters. Both these projects, DISNIC and AgRIS are under implementation in identified pilot districts.

However, the existing database on soil is inadequate to develop micro-level agricultural land use plan in the country, for which the needed scale of resolution is 1:4000/12500. The detailed digital database on soil (physical, chemical and biological) is a pre-requisite to address the various issues related to scientific Land Use Planning, soil reclamation; proper diagnosis of soils, judicious use of irrigation water and chemical fertilizers, nutrient deficiencies for maintenance of sound soil health and land productivity. The relevant soil parameters to be considered for such purposes are: soil type, elevation, type of land form, slope, geology (type of parent material), textural class, type of soil structure, soil water retentivity, soil pH, Electrical Conductivity (EC), Organic Carbon, CaCO3, Fe %, Major oxides, available macro and micro nutrients, depth of water table, erosion class, drainage & runoff characteristics, land capability and irrigability, etc. (Moni, M., 2011. Paradigm Shift in Geography: Challenges and Opportunities, a Keynote Address delivered at the National Symposium on Paradigm Shift in Geography organsied by A. M. Khwaja Chair, Department of Geography, Jamia Milia Islamia, New Delhi, on 28th November, 2011).

The National Mission on Sustainable Agriculture (NMSA) Report (2008, 2013) recommends undertaking Land Use Planning for Sustainable Agricultural and Rural development for effective production and productivity, and suggests creation of Natural Resources Management System using detailed soil resource mapping, and also Agricultural Resources Information System (AgRIS) at village level. To make Integrated Land Use Planning as a Decision Support System (DSS) tool, the Agricultural and Rural development Officers working at sub-district levels and District levels need to undergo both Capacity Building and Capability Building Programs during the 12th Plan period.

Agro Climatic Planning and Information Bank (APIB), covers agricultural land use/land cover; cropping systems; irrigated crop land; soils; degraded land; geomorphology; ground water potential; drainage characteristics cultural; features, agroecological characterization, database for agricultural practices, seeds, fertilizers, plant protection, agricultural implements, agricultural credits, insurance schemes, subsidy, agriculture market infrastructure, weather information, socio-economic infrastructure, sustainable agricultural land use plan based on integration of land capability, land productivity, soil suitability, terrain characteristics and socio-economic information using geospatial technology. In fact, there have been not many concerted efforts to revive farming or to develop new plans in the light of the results and conclusions of the micro-level studies. Moreover, the substantial gap existing between generated technologies and sensitiveness of agricultural technologies to the agro-ecological situations, social factors, economic factors and preferences of farm families have often been over-looked.

4.2 NEED OF THE HOUR

Convergence of various schemes at village and beneficiary level would lead to better outcome as seed improvement program, productivity enhancement program, promotion of organic farming, dry land horticulture, NHM, RKVY, MGNREGS, Minor irrigation department programs, etc. Funding is desirable to be continued central sector to evade diversion. There is a need to constitute a State Level Training and Monitoring Coordination Committee. Nodal officers at district, block and community need to be identified to monitor the progress and take corrective actions on a quarterly basis.

Establish linkages of all the institutions between all building the capacities of the grass roots to reach the large mass of farming community. Constitution of District level Training and Coordination Committee is necessary to guide planning, implementation and monitoring of land use plans. Adoption of cascading mode of training with the help of other training institutions and to develop the master trainers is essential.

The multi-layer, including remote sensing, databases need to be analyzed systematically to capture present land use and to evolve optimal land use plans. Geoinformatics tools and GIS platform may be fully internalized. The land not fit for agricultural and forest purposes can also be identified for industrial use.

5   GROUP MEETINGS

Ten groups (viz. Information needs on Natural resources for development of Land Use Plan (LUP), Analyzing Scientific and remote sensing data, Policy implication of scientific data bases on optimum land use plan, Legal frame work, Awareness and mobilizing farmer community, Network of training institutions and development of a cadre, Capacity building of officials; elected representatives and farming communities, Supply of scientific data bases for each Gram Panchayat/Village, Facilitating Land use planning and comprehensive plans and Convergence of Rural Development and Agriculture schemes and their monitoring) considered the needs of the hour of the workshop and made explicit recommendations including action initiating agencies. The details are in a separate chapter at the end.

CHAPTER 1

INTEGRATED LAND USE PLANNING FOR SUSTAINABLE AGRICULTURE AND RURAL DEVELOPMENT

M. V. RAO[1], V. SURESH BABU[2], K. SUMAN CHANDRA[3], and G. RAVINDRA CHARY[4]

CONTENTS

[1]Director General, National Institute of Rural Development and Panchayati Raj, Hyderabad – 500 030
[2]Associate Professor, National Institute of Rural Development and Panchayati Raj, Hyderabad – 500 030
[3]Professor and Head, Centre for Agrarian Studies, NIRD & PR, Hyderabad – 500 030
[4]Principal Scientist, CRIDA, Santhoshnagar, Hyderabad

1.1   INTRODUCTION

In India, sustainable development coupled with food security and poverty alleviation of rural poor is largely dependent on agriculture and allied activities as they support livelihoods of nearly 65.8 of India's rural population. In a developing country like India land is not only an important factor of production but also basic means of subsistence for majority of the people. Since Independence India's planners and policy makers have shown concern for efficient use of land, water and other natural resources for accelerated as well as economic development. The questions of efficiency, equity and environment protection have been flagged in almost all Five Year Plans. The market driven land use pattern may yield higher returns in the short run, but may pose several unmanageable problems for future generations due to unplanned overexploitation of land, water and other natural resources. The past and present experiences with unplanned land use, particularly in agriculture and rural development sectors resulted in unsustainable agricultural growth and production, enhanced land degradation leading to insecure livelihoods and poverty. These challenges necessitate for scientific land use planning and land use policy in all development sectors.

Land Use Planning (LUP) (FAO, 1993) aims at systematic assessment of land and water potential, alternatives for land use in relation to economic and social conditions in order to select and adopt the best land use options. Land use decisions are not made just on the basis of land suitability but also according to the demand for products and the extent to which the use of a particular area is critical for a particular purpose. Planning has to integrate information about the suitability of the land, the demands for alternative products or uses, and the opportunities for satisfying those demands on the available land, now and in the future. Therefore, LUP is not sectoral. Even where a particular plan is focused on one sector, for example, smallholder tea development or irrigation, an integrated approach has to be carried down the line from strategic planning at national, regional and local level.

1.2   LUP IN TERMS OF LAND, WATER AND OTHER RESOURCES: THE PRESENT STATUS

LUP aims at knowledge-based procedure that helps to integrate land, water, biodiversity, and environmental management (including input and output externalities) to meet rising demands of the human, livestock and other sectors, while sustaining ecosystem services and livelihoods. In a democratic country like India, LUP cannot be implemented with draconian measures. However, with the competing demands for growth and development in agriculture, rural development, industry, etc., various schemes and programs were initiated by central and state governments under the aegis of Planning Commission and tried to address either directly or indirectly the LUP in sectoral manner. For example, programs like Multi Cropping Program,

Command Area Development Program (CADP), Hill Area Development Program (HADP), Desert Development Program (DDP), Drought Prone Areas Program (DPAP), Whole Village Development Program, Decentralized Village Scheme, Integrated Waste Land Development Program (IWDP), Integrated Watershed Management Program (IWMP), Accelerated Rural Water Supply Program (ARWSP), accelerated Irrigation Benefits Program (AIBP) and Mahatma Gandhi National Rural Employment Guarantee Scheme (MGNREGS) of Ministry of Rural Development, GoI, Backward Region Grant Fund (BRGF) Program of Ministry of Panchayat Raj, GoI, and Comprehensive Land Development Program (CLDP) in Andhra Pradesh, etc. Many of these programs though have land based rural developmental activities but were implemented in a piece meal and top-down manner, many a times without considering the features of the locally existing natural resources (particularly land resources). As agricultural policies and water management strategies evolved over the years, water-use trends changed accordingly.

Land use planning is also integral to water resources development and management for agriculture, industry, drinking water, power generation, etc. In India, water resources development since independence significantly changed land use, particularly agricultural land use in Indo-Gangetic plains and below Deccan plateau. However, these changes could not bring sustained growth and development in agriculture due to externalities like mono-cropping; land degradation, equity issues, etc. The most popular and the most politically acceptable way of attempting to save water are to increase irrigation efficiency. However, International Water Management Institute's (IWMI) study has clearly shown that this method will not always be effective. Examining a hydrological system as a system of hydronomic zones has shown that efficient technologies will not be effective in natural and regulated recapture zones with groundwater storage and low salt buildup. Land and water must be managed in conjunction to achieve sustainable water use. A water-balance approach should be used to associate each land use with its associated net water depletion and to create a sustainable mosaic of land uses. For landscape-scale land-use planning, hydronomic zoning should be used to identify areas where improved irrigation efficiency would actually improve water management.

The late projects like "Sustainable Land, Water and Biodiversity Conservation and Management for Improved Livelihoods in Uttarakhand Watershed Sector," are aiming to improve the productive potential of natural resources and increase incomes of rural inhabitants in selected watersheds through socially inclusive, institutionally and environmentally sustainable approaches.

1.3 STATUS OF LAND RESOURCES INFORMATION

National Bureau of Soil Survey and Land Use Planning (NBSS&LUP) developed Soil Resource database on 1:250,000 at Association of Sub-Group level for broad planning at national and state level, soil resource maps for several districts

1:50,000 scale for district level land use studies, and selected large scale maps at 1:10,000/1:5000 at watershed/village level. The bureau also delineated 20 agro-ecological regions and 60 agro-ecological subregions (AESRs in the country) for agricultural planning. Spatial variability in the factors like bio-climate, physiography, soils and length of growing period were used in delineating the regions and subregions with details on agro-ecological settings and the land use potentials and constraints of each subregion along with information on major bench-mark soil series occurring in the sub region AESRs are homogeneous land units to implement a wide range of land resources applications. The agro-ecological zoning land resources appraisal linked to Geographic Information System (GIS) greatly enhanced the capability to develop alternate land use plan scenarios, management and decision support systems and improved interfaces to promote use of such systems together by scientists, development administrators and land users (Velayutham et al., 1999). Awareness about spatial variability in soil resources within microwatershed, enhanced productivity through crop/land use planning based on land capability and or land suitability was amply demonstrated in 54 micro watersheds across arid, semi-arid, irrigated, hill and mountain and coastal agro-ecosystems through NBSS&LUP led National Agricultural Technology Project (NATP) – Mission Mode Project on Land Use Planning for Management of Agricultural Resources, during 2001–2005.

In several locations only 50% of the farmers adopted land use that is considered scientifically correct or recommended by the NARS. One of the project outcomes distinctly brought out that in a microwatershed, cadastral level (1:10,000/25,000) both biophysical evaluation (soil resource inventory) and socioeconomic evaluation (Participatory Rural Appraisal, etc.) followed by allocating land parcels with appropriate land utilization types viz., crops, grasses, etc., on a topo-sequence, could enhance the land productivity from 20 to 50% compared to traditional land use across various soil types (NATP-MMLUP-Rainfed Agro ecosystem – Final Report, 2005).

1.4   LAND DEGRADATION

The information on the extent of soil degradation in the country has been assessed by various agencies and the estimates varied widely, that is, 63.9 m.ha. to 187.0 m.ha., due to different approaches and definitions in defining degraded soils and adopting various criteria for delineation. The main agencies that have estimated soil degradation are: National Commission on Agriculture (NCA, 1976), Society for Promotion of Wasteland Developments (SPWD, 1984), National Remote Sensing Agency (NRSA, 1985), Ministry of Agriculture (1985), and National Bureau of Soil Survey and Land Use Planning (NBSSLUP, 1984, 2005). The problems of land degradation are prevalent in many forms throughout the country. In most cases, a combination of such problems exists. In absence of comprehensive and periodic scientific surveys, estimates have been made on the basis of localized surveys and studies. NAAS-ICAR (2010) based on the harmonized area statistics of degraded and wastelands of India have given 120.72 million hectares. Still there needs to be a lot of clarity on the kinds and extent of the land degradation at local levels of administration to take up concerted efforts in agriculture and rural developmental activities by central and state governments.

1.5 LAND USE DATABASE

Out of the total geographical area of 328.73 million hectares, land use statistics are available for 305.61 million hectares, contributing 93% of the total. Till 1949–1950, the land area in India was classified into five categories known as the five-fold land-utilization classification. This five-fold land utilization classification was, however, a very broad outline of land-use in the country and was not found adequate enough to meet the needs of agricultural planning in the country. The states were also finding it difficult to present comparable data according to this classification, owing to the lack of uniformity in the definition and scope of classification covered by these five broad categories. To remove the noncomparability and to break up the broad categories into smaller constituents for better comprehension, the Technical Committee on Co-ordination of Agricultural Statistics, set up by the Ministry of Food and Agriculture (1948), recommended a nine-fold land-use classification replacing the old five-fold classification. The reliable land use data on various kinds and extent at various scales/levels is need of the hour. For example, the reliable data on Kharif fallows and Rabi fallows at any scale would be of much useful in land and water management, crop planning, policy decision making, etc.

1.6 AGRICULTURE LAND USE PLANNING: ISSUES

India has about 18% of world's population and 15% of livestock population to be supported from only 2% of geographical area and 1.5% of forest and pasture lands. The increasing human and animal population has reduced the availability of land over the decades. In recent years, land based livelihoods of small and marginal farmers are increasingly becoming unsustainable, since their lands do not support the family's food requirements and fodder for their cattle. As a result, rural households are forced to look at alternative means for supplementing their livelihoods. There is an estimate of declining agriculture landholding to 0.68 ha by 2020, and to a low of 0.32 ha in 2030. Out of 142 million hectares is net cultivated area, about 57 million ha (40%) is irrigated and the remaining 85 million ha. (60%) is rainfed. This area is generally subject to wind and water erosion and is in different stages of degradation due to intensive agricultural production. Therefore, it needs improvement in terms of its productivity per unit of land and per unit of water for optimum production.

More than 200 million of the rural poor live in the rainfed regions. These risk prone areas exhibit a wide variation and instability in yields. Diversification of land use is likely to be more in rainfed areas compared to irrigate and the contribution of rainfed agriculture would remain same at 44% to the total food grains. With about 68% of rural population (Kumar et al., 2009), these regions are also home to 81% of rural poor (Rao et al., 2005). Coarse cereals (85%), pulses (83%), oilseeds (70%), and cotton (65%) are the predominant rainfed crops grown in India (CRIDA, 2007). The cropping patterns have evolved based on the rainfall, length of the growing season and soil types. However, due to changed consumer preferences and market demand, farmers are now rapidly shifting to crops and cropping patterns which are more remunerative. The sharp increase in area under maize and cotton took place in few years at the cost of coarse cereals like sorghum, pearl millet mainly due to higher returns. Such changes will have implications on fodder availability to live-stock. This trend is likely to further increase by 2030.

The change in cropping pattern will have implications on the resource use. Con-tinuous monocropping increases vulnerability of farmers to weather risks, depletes soil fertility, ground water and leads to build up of pests and diseases. This issue has to be dealt through both technology and policy. For example, we have to evolve management practices for farmers' choice of remunerative crops without degrada-tion of the natural resource base and define agro-ecological zones where such crop-ping systems can be adopted in a sustainable manner. India has world's largest live-stock population accounting for about 55 and 16% of the world's buffalo and cattle populations, respectively. Rainfed areas account for two-thirds of total livestock population. India ranks second in goats (124.5 m out of world's 764.5 m) while third in sheep (59 m out of world's 1028.6 m), but of late there is a significant change in the livestock composition. India is globally the largest producer of milk with an an-nual production of 127.9 million tons during 2011–2012. Most of this production is coming from the rainfed regions where livestock population is large and is a major component of the livelihood systems. Currently the deficit in demand and supply of dry and green fodder stands at 12% and 63%, respectively and the gap is likely to be further widened with increase in demand for milk and meat products and change in cropping pattern like cotton and soybean at the cost of coarse cereals. There is a need to devote at least 10% of the net sown area for fodder production, which is presently around 5% (GoI, 2006).

India's natural ecosystems have immense richness of agricultural biodiversity including diversity in crops, plants, wild plants, livestock, aquatic species, below ground biota, microbes, etc. India has ranked seventh in the world in number of spe-cies contributing to agriculture and animal husbandry. Rainfed areas have a distinct superiority over irrigated in terms of diversity in term of flora and fauna. The entire forest area (76.9 m ha) is rich in biodiversity and it is rainfed. Apart from livestock and fishes, the contribution of forest areas to livelihoods of poor is substantial in rainfed regions. The forest area is major source of water and fodder. Climate change is likely to impact biodiversity in rainfed regions significantly. Rainfed areas show poor socioeconomic status, such as, limited irrigation facility, 15% as compared to 48% in irrigated regions, lower employment opportunities and higher population of agricultural labor force, lower productivity and lower per capita consumption, poverty, seasonal out-migration, poor infrastructure and social developmental indices.

Apart from above, population pressure, fragmentation of land holdings, tenancy farming, and low investment capacity, low productivity of crops and income, credit, pricing policy, marketing and wage rates are the major socioeconomic issues that impact rainfed agriculture. A major concern is the continuous decline in land holding size, which is a major hindrance in mechanization. Though significant proportion of the rural population is expected to move to cities by 2030, still substantial numbers will remain in rural areas. It is envisaged that leasing of the land and the pooling of the holdings may become more common in future. Urgent policy interventions are needed to protect the rights of the leaseholders so that they can invest on land development and soil improvement activities. Significant investments on secondary agriculture will be required to absorb the skilled labor force likely to be available due to the large investments of Govt. of India on skill development during XII Plan and beyond.

The issues are diverse and challenging in other agro-ecosystems viz, arid, irrigated, coastal and hill and mountain regions.

1.7   SPECIAL ECONOMIC ZONES (SEZS) LAND USE AND RURAL DEVELOPMENT: ISSUES

So far, 237 SEZs in 19 states out of which 63 have already been notified and the land to be acquired is estimated to be around 150,000 hectares of land, which is predominantly agricultural and typically multicropped producing close to one metric ton of food grain (Hiremath, 2007). If SEZs are seen to be successful in the future and if more cultivated land is acquired, this may affect country's food security. The issues regarding SEZs vis-a-vis agricultural land use and rural development are of major concern today and need to be discussed in a food, nutritional and rural livelihood security context.

1.8   LAND USE FOR RURAL DEVELOPMENT (NON-AGRICULTURE): ISSUES

Many rural infrastructure programs like rural connectivity, housing, sanitation and health, drinking water, rural institutions, etc., are land based activities and need land, preferably non-arable and low or no productive lands. The decisions for allocation of land for such basic needs have to be taken at PRIs level. In view of the locally existing sociopolitical-economic dynamics, taking such decisions in a bottom up and comprehensive manner needs a fairly good understanding of the land resources and their planning for multifarious activities. Thus, there is a need to develop information on land resources as a database at local level in an understandable manner.

1.9   INTEGRATED LAND USE PLANNING AND SUSTAINABLE AGRICULTURE AND RURAL DEVELOPMENT (SARD): PROCESS, ELEMENTS AND MAJOR ISSUES

The Sustainable Agriculture and Rural Development (SARD) Initiative is a multi-stakeholder umbrella framework that engages civil society, governments and inter-governmental organizations in a joint effort to make rapid progress toward achievement of the Agenda 21 vision of the Earth Summit (1992) for SARD. The SARD Initiative emerged from the Dialogue on Land and Agriculture at the Eighth Session of the UN Commission on Sustainable Development (CSD-8) in 2000 and the subsequent SARD Forum that was organized as a side event at the FAO Committee on Agriculture (COAG) in 2001. In the run-up to the World Summit on Sustainable Development (WSSD), CSD decided to give much greater emphasis to implementation processes involving stakeholder partnerships. FAO, in its capacity as Task Manager for Chapter 14, facilitated preparation of the SARD Initiative, with active participation of civil society.

FAO defines SARD (Sustainable Agriculture and Rural Development) as a process, which meets the following criteria:

- ensures that the basic nutritional requirements of present and future generations, qualitatively and quantitatively, are met while providing a number of other agricultural products.
- provides durable employment, sufficient income, and decent living and working conditions for all those engaged in agricultural production.
- maintains and, where possible, enhances the productive capacity of the natural resource base as a whole, and the regenerative capacity of renewable resources, without disrupting the functioning of basic ecological cycles and natural balances, destroying the sociocultural attributes of rural communities, or causing contamination of the environment.
- reduces the vulnerability of the agricultural sector to adverse natural and socioeconomic factors and other risks, and strengthens self-reliance. (From FAO Trainer's Manual, Vol. 1, "Sustainability issues in agricultural and rural development policies," 1995)

Elements for Sustainable Agriculture and Rural Development – Integrated Activities

- Government level – Policies, instruments, development plans, agrarian reforms, untraditional surveys, food quality and food security, data, monitoring, early warning systems.
- Rural community level – Development of local organization and capacity building for people's participation, training, extension.
- Area level – For example, coastal zones, watersheds, river basins, agro-ecological zones.

- Production unit level – Farming systems, diversification to increase incomes, creation of rural industries, credit and marketing.
- Consumer level – Improving nutrition and food quality, adjusting dietary patterns, product marketing, and consumer forums for conflict resolution.

1.10 MAJOR ISSUES/PROGRAM FOR DISCUSSION ON ILUP FOR SARD

In view of the emerging issues in LUP for SARD and considering the issues of the multifunctionality of land and land use decisions, scale and data requirement, data integration, and changing concepts and tools for LUP, major issues related to technical, policy and support systems, legal and capacity building for furthering scientific land use planning are highlighted in the following sections, in which some were also highlighted in the Agenda Note of the National Brain Storming Workshop organized by ICAR at NBSSLUP during 25–26 July 2008 at Nagpur.

1.10.1 TECHNOLOGICAL ISSUES

- Land use data generation at cadastral level with details about land tenure and land ownership to make effective land use plans. There is need to relook to resurvey and reclassify the land records and data collection mechanism at grass root level.
- The exploratory DSS and its importance in data integration and analysis in LUP processes to provide answers for how, where, when to grow different crops to enhance the food production in the country.
- LUP is either prescriptive or exploratory at regional (state/district) level, which provides a window for development options while LUP becomes increasingly participatory at village and watershed level advocating certain land uses at local level and also takes care of self-interest of the landholder.
- Improving farm production and farming systems through diversification of farm and nonfarm employment and infrastructure development

1.10.2 POLICY ISSUES

- Land use planning has geographical aspects like geology, soils, and climate, socioeconomic aspects, which need to be synchronized in data integration processes. Land resource should be used as per its ecological capability.
- Agriculture policy review, planning and integrated programming in the light of the multifunctional aspect of agriculture, particularly with regard to food security and sustainable development

- State Land Use Boards have to be made more active and effective in the land use planning processes at state level in general, and agricultural plans.
- Various policies and their importance should have harmony among the forest, agriculture and environment, which is possible through integrated approach to achieve the ecological security of the country.
- The district level land use plans are good enough to capture the intradistrict and interregional variability in resources availability.
- Land use in general and land use planning in particular is closely linked to land tenure, land title, land rights, land lease issues, etc.
- Land Conservation and Rehabilitation
- Water for sustainable agriculture and sustainable rural development

1.10.3   SUPPORT SYSTEMS OPTIONS

- Estimating people's participation and promoting human resource development for sustainable agriculture
- Market mechanism operating more strongly than in the past, crop and land use shifts will become more dynamic now.
- Land use changes and LUP should take care the rainfall aberrations, soil health, market forces and nutritional requirements.
- New tools like crop or weather insurance can also be used as incentives to encourage scientific land use in rainfed areas.
- New opportunities are also arising in the area of clean Development Mechanism (CDM) and carbon credits where farmers can be compensated for adopting conservation practices which contribute to scientific land use and sustainable productivity on a long-term basis, but relatively lower returns on short-term.
- A combination of better technologies, new policies and support systems are required to meet the goal of realizing higher productivity and simultaneously adopting a sustainable land use.

1.10.4   LEGAL ISSUES

- In view of the land use, which was perceived as a matter of local concern, has become a matter of national concern and hence the basis of distribution of power in relation to land, which was relevant and valid when constitution was drafted, has become outdated. Therefore, the amendments of the constitution transferring the subject of land from state list to the concurrent list (Komawar and Deshpande, 2008) needs due consideration.

1.10.5   CAPACITY BUILDING

- Land resource planning information and education for agriculture and rural development for all stakeholders

Keeping the above-issue in view, a national level workshop was organized at National Institute of Rural Development and Panchayati Raj (NIRD & PR), Hyderabad with the following objectives:

1. to examine the database (situation analysis) on natural resources with a view to identify major constraints/challenges in Land Use Planning (LUP) with focus on SARD;
2. discuss the strategies and programs for LUP to address SARD;
3. identify policy related issues for SARD; and
4. suggest mechanisms for integrating RDPs with LUP and also institutional arrangements.

1.11   THE WORKSHOP CREATED A PLATFORM TO ADDRESS CERTAIN ISSUES

- What are the Sources, scales and quality of land resources and land use data for efficient planning?
- Which are the hot spots in the agro-ecological systems that are degraded? How to rejuvenate such areas in terms of quality?
- What should be the unit of planning for holistic approach to conserve and efficient use of natural resources at different administrative levels/domains?
- What is the role of state land use board, district planning committees and its implications on LUP?
- How to integrate various programs and schemes implemented by central and State to address SARD?
- What is the need, role and the capacities of the Gram Panchayat members for holistic planning and enhance the manpower at GP level for efficient planning?
- What are the technologies, new policies and support systems required to address SARD?
- What are the Legal issues relevant to integrate LUP?
- What should be role of community in planning for poverty alleviation and integrated planning for SARD and support NRM based livelihoods?
- Constraints in preparing GIS based planning and at what level GIS based planning is accurate and possible for integrated land use planning for SARD?

The papers presented in the workshop by the academicians and researchers are brought out as an edited volume for the benefits of research scholars and policy makers and all stakeholders.

KEYWORDS

- **Agriculture Land Use Planning**
- **ILUP**
- **Land Degradation**
- **Land Use Database**
- **LUP**
- **Rural Development**
- **SARD**
- **Special Economic Zones**

REFERENCES

FAO. (1976). A Framework for land evaluation. Soils Bulletin 32, FAO, Rome.72 p. Also, Publication 22, (R. Brinkman and A. Young (eds.), ILRI, Wageningen, The Netherlands.

FAO. (1993). Guidelines for Land Use Planning. FAO Development Series 1. FAO, Rome. pp. 231.

Katar Singh, (2009). Rural Development – Principles, Policies and Management. Third edition. SAGE Publications India Pvt. Ltd., New Delhi.

Komawar and Deshpande, (2008). In: Concept Notes. Legal Aspects of Land Use Planning and Policy. National Brain Storming Session on Land Use Planning and Policy Issues at NBSS&LUP (ICAR), 25–26 July 2008, Nagpur.

NAAS-ICAR, (2010). Degraded And Waste Lands Of India – Status and Spatial Distribution, Indian council of Agricultural Research, New Delhi. Published by Directorate of Information and Publications of Agriculture, ICAR New Delhi. pp. 167.

NATP-MMLUP-Rainfed Agroecosystem – Final Report, (2005). NATP-MMLUP-III-28-Land Use Planning for Management of Agricultural Resources.-Rainfed Agroecosystem, Central Research Institute for Dryland Agriculture, Hyderabad. pp. 105.

NBSS & LUP, (2008). Agenda Note. National Brain Storming Session on Land Use Planning and Policy Issues at NBSS&LUP (ICAR), 25–26 July 2008, Nagpur.

NSSO (2009–2010), Employment and Unemployment survey unit level data.

Ravindra Chary, G, Vittal, K. P. R, Ramakrishna, Y. S., Sankar, G. R. M., Arunachalam, M., Srijaya, T., Udaya Bhanu. (2005). Delineation of Soil Conservation Units (SCUs), Soil Quality Units (SQUs) and Land Management Units (LMUs) for Land Resource Appraisal and Management in Rainfed Agroecosystem of India—A Conceptual Approach. Lead Paper. Proceedings of National Seminar on Land Resources Appraisal and Management for Food Security, organized by Ind. Soc. of Soil Survey and Land Use Planning, NBSSLUP, Nagpur, 10–11, April, 2005.

United Nations General Assembly, (2002). The Johannesburg Declaration on Sustainable Development, United Nations.

United Nations General Assembly. (1992). Rio Declaration on Environment and Development, United Nations.

Velayutham, M., Mandal, D. K., Mandal, Champa, Sehgal, J. (1999). Agro-ecological Sub-regions of India for Planning and Development. NBSS Publ. 35, 372 p. NBSS & LUP, Nagpur, India.

World Bank, (2001). Making Sustainable Commitments: An Environmental Strategy for the World Bank, http://www.worldbank.org/environment/.

CHAPTER 2

SOIL AND LAND USE SURVEY FOR STATE LEVEL PERSPECTIVE LAND USE PLANNING AND PRAGMATIC FARM LEVEL PLANNING

M. VELAYUTHAM[1]

CONTENTS

[1]Former Director, National Bureau of Soil Survey and Land Use Planning (ICAR), and Former Executive Director, M.S. Swaminathan Research Foundation.

2.1   SOIL VARIABILITY AND SOIL SURVEY

The earliest investigation on soils of India date back to Leather (1898) who categorized the soils of India into four major soil groups viz. Indo-Gangetic Alluvium, Black cotton soil, Red soil and Laterite soil, which is now in common knowledge with the farmers and also easily recognized by the people. Moreover, the immense variability and complexity of soil behavior is also perceived by practicing farmers who recognize the differential response of the land to soil management and production inputs between different soil types.

In his opening address at the meeting of the Consultative Group for International Agricultural Research (CGIAR) held at New Delhi in 1994, the late Prime Minister Shri. P.V. Narasimha Rao stated: "Each plot of land is like a human being. It has to be tended like a child. There is a need to find differentiated and properly considered prescription for each of these varieties rather than tarring with one kind of brush, which is not going to work in agriculture. I can tell you the characteristics of each and every survey number which I own in my village, because I have seen it yield, failing to yield and under some conditions it refuses to yield. So far, we have only been working in agriculture on general prescriptions. From the general to specific technologies is a long journey to be undertaken. So far, we have tried to produce what we need by hook and crook, by getting hold of the best land, best inputs and best of everything. The Green revolution methods are going to be found inadequate hereafter. There is a need to diagnose the nature, properties, potential and problems of each parcel of land."

Specific soil survey investigation during the early twentieth century resulted in the production of soil maps for specific areas and for specific purposes. Tamhane (1965) highlighted the importance of soil classification, soil surveys and soil interpretation and their utilization to cater to the needs of National agricultural programs. Raychaudhuri (1968) proposed 28 broad soil classes and suggested that composite map of each village be prepared in the scale 1:7920 or 1:3960 giving boundaries of land use classes and cropping pattern. Govindarajan (1971) produced a soil map of India in 1:7 million scale depicting 23 major soil groups. Raychaudhuri and Govindarajan (1971) detailed the nature and properties of the soils of India in the ICAR technical bulletin (AGR No. 25) entitled "Soils of India."

During the period 1986–1999, the National Bureau of Soil Survey and Land Use Planning (NBSS & LUP) of the ICAR completed a nation-wide soil survey at 10 Km grid interval and brought out soil map of different states in 1: 250,000 scale. Velayutham et al. (1999) using the soil, land and climatic features, delineated India into 60 agro-ecological subregions for developmental planning. Bhattacharyya et al. (2013) compiled and presented the soil information from the above-mentioned project at the level of Family Associations following the USDA soil Taxonomy. Out of 7 soil orders and 11 major soil groups, 1247 soil families have been identified. The 1:250,000 scale map also shows a threshold soil variation index of 4–5 and

10–25 soil families per million hectare for alluvial plains and black soil regions, respectively.

The soil survey data interpreted for land capability classification can be used for land use planning at hierarchical levels depending on the intensity of survey undertaken and scale of mapping. Dovetailing soil survey data with other natural resources data and socioeconomic parameters, macrolevel planning of land use for different sectors can be prepared at levels of agro-ecological regions, Agro-ecological subregions and agro-ecological zones at state level for development planning.

2.2   INTERPRETATIVE USE OF SOIL SURVEY INFORMATION

At operational level the soil survey data are useful at watersheds, irrigation command arcas and farm levels for the following purposes:

1.   crop planning under rainfed agriculture;
2.   depth and frequency of irrigation in command area;
3.   drainage requirements assessment;
4.   evaluation of crop productivity potential of classified soil units – to bring out the differences in intrinsic soil properties that influence crop productivity at the same level of management;
5.   use of soil survey information for soil erosion assessment and prioritizing soil conservation areas and measures;
6.   soil based agro-technology transfer based on the characterization of soils of research stations to similar soils in the area by delineating extension domain based on benchmark soil concept;
7.   soil survey information with other natural resources and socioeconomic factors in GIS framework for developing integrated land use and management predictions and validation.

The National Commission on Agriculture (1972) and the task forces constituted by the Planning Commission in 1972 and 1984 have emphasized the need for systematic and detailed soil survey of the agricultural lands at large scale mapping for scientific crop and land use planning.

2.3   DEVELOPMENT OF SOIL PRODUCTIVITY INDICES AND RATING

The earliest approach for such an exercise was initiated by Shome and Raychaudhuri in 1960. Using the following three factors, they rated the soils in 294 districts for their production potential:

1.   character of soil profile;
2.   topography, texture and structure;
3.   (i) degree of climatic suitability; (ii) salinity; (iii) stoniness; (iv) tendency to erode.

Based on the production potential field experiments under irrigated agriculture and rainfed agriculture conducted by the All India coordinated research projects of the ICAR and SAUs and the soil survey information available for the different districts, there is now a need to assess again the production potential and rating of various soil classes mapped at state and district level. Such an exercise along with detailed soil survey available at 1:10,000 scale using cadastral map as the base will lead to the preparation of village level/watershed level soil maps with land management units delineated for adoption of appropriate soil and crop management practices. A case study based on this approach for the soils of Andhra Pradesh has been brought out by Naidu et al. (2013).

2.4   DETAILED SOIL SURVEY FOR FARM LEVEL LAND USE PLANNING

As indicated earlier the immense variability in the properties of the soils including the fertility parameters of the surface soils necessitate the need for the detailed soil survey for watershed level/farm level land use planning and land management operations. Based on the project proposal entitled "Establishing Land resource Database for farm planning in Tamil Nadu," submitted by the state level task force under my chairmanship in 2003, the pilot project for detailed soil survey was jointly undertaken in 17 blocks by the government of Tamil Nadu, Agricultural University and the Bangalore Regional Centre of the National Bureau of Soil Survey and Land Use Planning. This experience has provided the methodological basis for detailed soil survey and land resource mapping for farm level and village level land use planning as detailed by Velayutham and Natarajan (2003). This paradigm shift from soil mapping to land resource mapping at village level by detailed soil survey and socioeconomic survey is suggested for embarking upon a National Mission Mode Project, which will help in the development of soil productivity indices and rating, land capability classification and extent of degraded lands at village level and development of a National Portal on soils of India for knowledge based agro-technology transfer by modern methods of ICT for ensuring sustainable agriculture and food security for the people now and in the future, Velayutham (2012).

KEYWORDS

- **Agro-technology**
- **Black cotton soil**
- **Indo-Gangetic Alluvium**
- **Laterite soil**
- **Red soil**
- **Soils of India**

REFERENCES

Bhattarcharyya, T., et al. (2013). Soils of India; Historical Perspective, Classification and Recent Advances. Current Science 104: 10: 1308–1323.

Govindarajan S. V. (1971). Soil Map of India 1: 7 million Scale In Review of Soil Research in India. Kanwar, J. S., Raychaudhuri, S. P. (eds.)

Leather (1898). In Agriculture Ledger. Composition of Indian Soils.

Naidu et al. Agropedology. 23 (1):30-35

Naidu et al. (2013). Need for Management Domain Soil Maps: A Case study of Andhra Pradesh. Agropedology. 23 (1):30-35.

NCA (1972). Interim Report on Soil Survey and Soil Map of India, National Commission on Agriculture.

NNRMS (1984). Report of the Task Force on Soil and Land Use, National Natural Resources Management Systems, ISRO, Bangalore.

Raychaudhuri, S. P. (1968). Interpretative Soil Groupings – Prediction of Soil Behavior and Practical use of Soil Maps, J. Indian Soc. Soil Science, 205–213.

Raychaudhuri, S. P., Govindarajan, S. V. (1971). Soils of India. ICAR Tech. Bull. (AGR) No.25, p.45.

Shome, K. B., Raychaudhuri, S. P. (1960). Rating of Soils of India. In Proc. Natl. Institute of Sciences, India 26 Part A Physical Sciences, Supplement I-II, 260 289.

Tamhane, R. V. (1965). Soil Investigations – Their Orientation to the Needs of a National Agricultural Program, J. Indian Soc. Soil Science, 211–215.

Velayuham et al. (1999). Agro-Ecological sub-regions of India for Development Planning, NBSS & LUP, Nagpur, Pubn. N. 35, 452.

Velayutham, M. (2012). National Soil Information System (NASIS) and Land Resource Mapping for Perspective Land Use Planning and Pragmatic Farm Level Planning, Madras Agric. J., 99 (4 6), 147–154.

Velayuhm, M., and Natarajan, A.(2003). Task Force Report: Project proposal for establishing Land Resource Database for Farm Planning in Tamil Nadu. Dept. of Agriculture, Govt. of Tamil Nadu. 17 p.

PART I

SOIL AND LAND RESOURCE INFORMATION AND LAND DEGRADATION ASSESSMENT

CHAPTER 3

NEED FOR CADASTRAL LEVEL LAND RESOURCE INFORMATION FOR SUSTAINABLE DEVELOPMENT—A CASE STUDY IN CHIKKARASINAKERE HOBLI, MANDYA DISTRICT, KARNATAKA

A. NATARAJAN, RAJENDRA HEGDE, R. S. MEENA, K. V. NIRANJANA, L. G. K. NAIDU, and DIPAK SARKAR[1]

CONTENTS

[1]National Bureau of Soil Survey and Land Use Planning, Regional Centre, Hebbal, Bangalore-560024, India.

ABSTRACT

The scientific prerequisite for implementing, monitoring, reviewing and evaluating all the land developmental programs is the generation of detailed land resources database. This can be obtained only by conducting cadastral level detailed surveys wherein we characterize and delineate homogenous areas based on soil-site characteristics into management units. The database serves as a long-term aid for all the land based development programs and it needs only occasional checks. Detailed cadastral level survey taken up to study the status of soil and other resources occurring in Chikkarasinakere block of Mandya district in Karnataka during 2010 has brought out the farm-wise status of land resources and revealed that land degradation is at an alarming stage. Nearly 59% of the area is suffering from various degrees of chemical and physical degradation. The situation becomes very alarming because the soils of the area were well drained red soils highly suitable for irrigated agriculture when irrigation was introduced during 1905. The process of degradation shall attain faster phase if appropriate interventions/investments are not undertaken on priority. Continuation of present management practices can damage the soil health very rapidly. As the command area is one of the important rice bowls of Karnataka, there is an urgent need to reverse the process of degradation by adopting site specific interventions as indicated in the study. The study demonstrates the immense utility of detailed land resources database.

3.1   INTRODUCTION

The urgent need for conserving our soil and water resources is felt more at present than ever before, because of the fast degrading natural resource base and stagnating crop yield levels. The increase in productivity to meet the demands of growing population and to ensure delivery of food grains as per the terms of proposed Food Security Bill, India needs an additional production of 70 million tons per year. This is possible only by targeting the potential areas and overcoming the constraints in marginal lands. According to the estimates available from various agencies, about 120 m.ha is affected by land degradation, 55 m.ha area lying as wastelands, 10% of the irrigated lands affected by salinity and alkalinity, widespread deficiency of secondary and micronutrients, over exploitation of ground water, increasing area under current fallows, diversion of prime lands to non agricultural uses and declining factor productivity threaten the food security of the country.

Creation of irrigation facility has resulted in rural prosperity in many parts of India by increasing the crop productivity and cropping intensity. One of the important inputs responsible for "Green Revolution" is the provision of irrigation. Of late, the crop productivity levels in many irrigated command areas have plateaued or started declining rapidly due to the deterioration of soil health. Unscientific and excessive irrigation, growing crops not suitable to the soils and unscientific management of soils are the main causes for the present situation. Waterlogging, increased salinity/ sodicity, nutrient imbalance, shrinking diversity of microflora and fauna have become major constraints limiting the crop choice and crop productivity

It is obvious that the neglect and deterioration of the land resources and consequent decline in the productivity is continuing without any check for many years in the country. The consequence of this neglect is already expressed in many forms and its impact will be felt severely on the environment and economy of the nation in the future. The existing situation is very much likely to worsen in the years to come and warrants urgent course correction. Another disturbing fact is that almost all the above estimates are based on sample survey only.

Government of India spends every year nearly one lakh crore rupees on various agricultural development programs, soil and water conservation being the most prominent activity. However, various post project evaluation reports suggest that 50–58% of farm-land areas are suffering from various kinds of degradation. It indicates that there is mismatch between the actual requirement and what is actually implemented in the field in the form of various land based rural development programs. Such situations can be avoided if detailed, farm specific land resources database are generated in each and every project area to identify inherent potentials and constraints and evaluated for suitability to various land uses.

It is well known that the root cause for the degradation, neglect and irrational use of land resources exists at the grassroots level, at the doorstep of each and every farmer, living in thousands of villages in the country. For this, the first and foremost thing needed is a detailed site-specific database on land resources at the farm level for all the villages in the country. To prove the importance of farm level information for sustainable resource management, a case study was done at Chikkarasinakcre Hobli, in Mandya District, of Karnataka during 2010–2011. Like in any other Hoblis in the country, the uplands are facing problems like soil erosion, nutrient loss, and the lowlands showing declining factor productivity, water logging and salinity. Though these problems are known in general to the developmental agencies of the area, the exact nature, extent and severity are not brought out clearly by the earlier datasets generated from the reconnaissance and other small-scale surveys. To fill this gap and know the exact nature and severity of salinization in the irrigated areas and type of degradation in the upland soils, detailed cadastral level inventory of soil and other land resources in the area was taken up.

3.2   GENERAL DESCRIPTION OF THE AREA

Chikkarasinakere Hobli, with an area of 16,873 ha, is located in the eastern part of Mandya district. This Hobli has 42 revenue villages (Fig. 3.1). Major part of the area (7478 ha) is under canal irrigation and rainfed area occupies about 4367 ha. Granites and gneiss are the major rock types of the block. Dykes and pegmatite and quartz veins occur in few places in the area. The granites, gneiss, and dykes with their widely varying composition and rate of weathering have influenced significantly the formation or genesis of different type of soils observed in the block.

3.2.1   LAND FORM ANALYSIS

The landscape of the area has evolved over a period of time due to the process of erosion and deposition of weathered rocks, minerals, soil and other materials from one place to another. Uplands, lowlands and valleys are the major landforms identified in the area. The uplands are characterized by nearly level to very gently sloping summits, followed by gently to very gently sloping side slopes, which merges with nearly level lowlands/valleys. The gently to very gently sloping side slopes (midlands) occur extensively in all the villages (Photograph 3.1). The lowlands occur mostly below the tanks or between the uplands and valleys. They have flat topography and mostly cultivated to paddy or sugarcane (Photograph 3.2). The uplands occupy about 55% and the valleys and lowlands occupy about 29% of the area.

FIGURE 3.1   Location of Villages in Chikkarsinakere Hobli.

PHOTOGRAPH 3.1 Very Gently Sloping and Bunded Uplands in S.No. 50, Arechakanahalli Village.

PHOTOGRAPH 3.2 Nearly Level to Level Lowlands Near the Stream in Madenahalli Village (S.No. 6 and 7).

3.2.2   SLOPE DETAILS

The elevation ranges from 600 to 680 m AMSL. The general slope of the block is from west to east direction. The slope of the land surface has played a critical role in the formation of soils in the area. Generally it has been observed that the depth of the soil formed is directly proportional to the length of the slope. The block is drained by Shimsha River and many small streams and nallas. All the streams and nallas are seasonal in nature and only the main river has flow throughout the year. The introduction of irrigation from K.R. Sagar during 1905 has completely changed the earlier drainage pattern in the block. At present, the block is irrigated by two main branches of Viswesvaraya canal. These two main branch canals with their network of numerous channels and field outlets link all the villages and hamlets in the block in an intricate manner. Due to this network, the area is supplied with Cauvery water almost throughout the year. This has created serious drainage problem, particularly in the low-lying areas of the block.

3.2.3   CLIMATE

The Chikkarasinakere Hobli enjoys subtropical monsoon climate. The average temperature of the area ranges between 16°C and 35°C. The normal rainfall received is about 770 mm. Out of this, about 50% is received during south-west monsoon, 20% during north-east monsoon and 30% during the summer period. The average number of rainy days during this eight-month period is about 43 days. The total potential evapo-transpiration is about 1794 mm, which is very much higher than the amount of rainfall received in the area. Only during September and October months the precipitation is more than the PET. Whenever the rainfall reaches nearer to the ½ PET values, that period is considered favorable for plant growth. This period gets extended to one or two weeks more if the soil conditions are favorable (soil depth, texture, structure, high organic matter contend, etc.) for plant growth. Accordingly the growing period for the area is calculated as 150 days, which starts from the middle of July (28th week to 50th week) and extends up to the middle of December (Fig. 3.2). The growing period determines the choice of crops, particularly in the rainfed areas of the block.

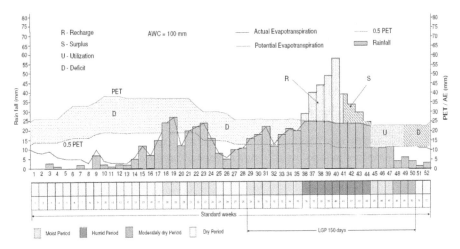

FIGURE 3.2 Rainfall, PET and Growing Period of Chikkarasinakere Hobli.

PHOTOGRAPH 3.3 Sugarcane Planting on Ridges in Madarahalli Village.

3.3 METHODOLOGY FOLLOWED FOR CADASTRAL LEVEL RESOURCE CHARACTERIZATION

The purpose of this cadastral level survey was to delineate similar areas, which respond or expected similarly to a given level of management. This is achieved in the field by detailed study of all the site (like slope of the land, erosion, drainage, salinity, occurrence of rock fragments, etc.) and soil characteristics (like depth, texture, color, structure, consistency, coarse fragments, porosity, soil reaction, etc.), grouping similar areas based on soil-site characteristics into management units and showing their extent and distribution on a village cadastral map.

3.4   FIELD INVESTIGATIONS

The detailed survey of 42 villages in the Hobli was carried out by using cadastral map as a base in conjunction with remote sensing data products (IRS P6, LISS IV MX) at 1:12,500 scale. After identifying the landforms and surface features and delineating them on the base map, preliminary traverse of the Hobli was carried out. During the traverse, major rock types of the area, drainage pattern, surface features like erosion, rock out crops, presence of stones and gravels at the surface, etc., slope characteristics, existing land use and landforms were identified and soils studied in few places. Based on the above information collected and by referring the earlier soil survey information available for the area, an initial legend showing the tentative relationship that exists between the geology of the area, landforms identified and major soils occurring in the area was prepared (Natarajan and Dipak Sarkar, 2010; Soil Survey Staff, 1951; Soil Survey Staff, 1993; Soil Survey Staff, 1996).

3.5   RESOURCE MAPPING AT VILLAGE LEVEL

After establishing the soil-landform relationship for the block, detailed characterization of soil, site and other parameters were carried out for each village separately. In each village, the field boundaries and survey numbers given on the cadaral sheets were located on ground by following permanent features like roads, railway lines, nallas, streams, rivers, tanks, ponds, burial ground, etc., and wherever changes were noticed they were incorporated on the cadastral maps. After this, intensive traversing of each physiographic unit like ridges, uplands, lowlands/valleys, etc., was carried out by involving all the soil survey party members. Based on the variability observed on the surface, transects were selected across the slope, covering all the physiographic units occurring in each village.

In the selected transect, profiles were located at closely spaced intervals to take care of any change in the land features like break in slope, erosion, gravel, stones, etc. At the selected sites, profiles were opened up to 200 cm and studied in detail for all their morphological and physical characteristics. Based on the soil site characteristics, the soils were grouped into different soil series. Soil depth, texture, color, amount and nature of gravel present, calcareousness, presence of limestone, nature of substratum and horizon sequence were the major identifying characteristics of soil series in the area surveyed. Based on the above characteristics 13 soil series were identified in the Hobli. The soil series were further divided into 118 phases (phase is a subdivision of a soil series), which are based mostly on surface features surface like texture, slope, erosion, presence of gravels, salinity, sodicity, etc. that affect its use and management. For the series identified, soil samples were collected from representative pedons for laboratory characterization (Jackson, 1973).

3.6   FINALIZATION OF SOIL MAPS

The soil map for each village was finalized in the field after thorough checking of soil and site characteristics and correlation. The soil map thus prepared shows the individual field boundaries, their survey number and various soil or management units (showing the phases of soil series) occurring in the surveyed area (Fig. 3.3). Out of the 13 soil series mapped eight series occur in uplands, one series in midlands and four series in lowlands (Photographs 3.4–3.6). The generalized soil map of the block was prepared by combining all the soil maps of the villages.

3.7   GENERATION OF THEMATIC MAPS

The soil-site characteristics were interpreted for identifying the constraints and potentials like soil depth, erosion, gravelliness, calcareousness, salinity and land suitability for various crops and other uses for each village separately by using GIS software.

PHOTOGRAPH 3.4   Kulagere series, S.No. 13, Aravanahalli, Yadaganahalli series, Kulagere Panchayat.

PHOTOGRAPH 3.5   Profile Features of S.No. 38, Yadaganahalli village.

PHOTOGRAPH 3.6   Soil Characteristics of Chikkarasinakere series, S.No. 376, Kulagere village.

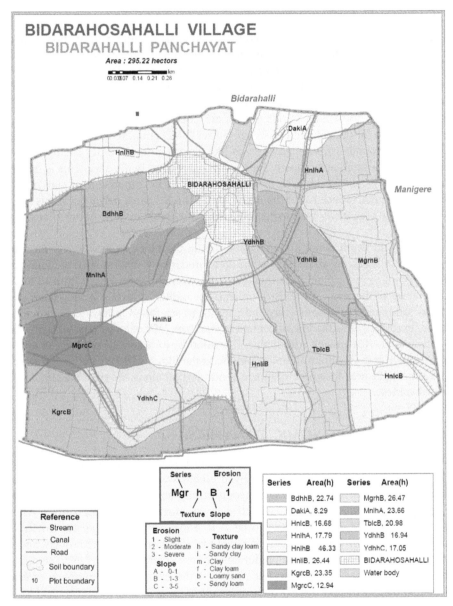

BIDARAHOSAHALLI VILLAGE
BIDARAHALLI PANCHAYAT
Area : 295.22 hectors

FIGURE 3.3 Soil Map of Bidarahosahalli Village Showing the Distribution of Management Units.

3.7.1   LAND USE

Out of the total area of 16,873 ha, about 12,739 ha area is under cultivation, which is more than 75% of the total area available in the block. Forests are almost nonexistent in the area. Barren and uncultivable lands occupy very less area. Though the land put to nonagricultural use is less, it shows an increasing trend. Major parts of the cultivable lands (7478 ha) are under canal irrigation. The canal-irrigated lands are used for the cultivation of rice, sugarcane, coconut, mulberry, etc., and rainfed areas are used for the cultivation of ragi, pulses (blackgram, redgram) and oilseeds (sunflower, groundnut). Vegetables crops like tomato, brinjal, chilies and cucumber are grown in a small area, mostly for local consumption in the area. This is one of the intensively cultivated blocks in Mandya district. The importance of tank irrigation is very much reduced after the introduction of KRS canal irrigation in this area. Similarly, the area under well irrigation is very negligible in most of the villages in the block.

Soils: Soil depth is not a major limitation in the block as moderately deep to very deep soils occupy about 68% of the area. Moderately shallow soils are found in about 15% area in the block. The thickness of surface soil layer varied from 13 to 20 cm in most of the area. The soils of uplands are lighter in texture (sandy loam and loamy sand) and heavy soils (clay and clay loam) are found in low lands and midlands. In surface soils the gravel content is less than 25%, whereas subsurface soils contained 30–70% gravel. Although soil erosion is not a serious problem in about 82% area, it is a constraint in upland areas. In the block, 53% area is well-drained and about 11% area face poor drainage conditions all along the river Kadambi and its tributaries.

3.7.2   SOCIO-ECONOMIC DYNAMICS

With gently sloping landscapes, deep well drained loamy to clayey soils, good and fairly well distributed rainfall, interspersed with network of tanks and many small and medium streams, the Hobli is a predominantly an agrarian area of the district. Prior to the introduction of canal irrigation it was mostly under rainfed crops. After the introduction of canal irrigation from K.R. Sagar during 1905 all the lowlands and most part of the very gently sloping uplands, after terracing, were brought under irrigation. Due to this, the once predominantly dry rainfed tract has become a wet zone and consequently there was a total shift in the cropping pattern practiced earlier, dominated by ragi to paddy and sugarcane at present. Consequently, the Hobli has become one of the most intensively cultivated and comparatively prosperous regions of Mandya district.

3.7.3   PRESENT STATUS OF LAND RESOURCES

Canal irrigation and associated factors have brought in a dramatic change in the landscape of the area, with lush green paddy and sugarcane fields, interspersed with coconut and mulberry gardens seen everywhere in the Hobli. But the large-scale intensive paddy and sugarcane cultivation year after year during the last 100 years has brought in very significant change in the quality and productivity of soil, water and other resource endowments. The impact is very visible, particularly on the soil resources of the area. Continuous irrigation has changed significantly the drainage pattern, soil morphology, altered or destroyed the soil structure, physical properties and chemical and biological composition of the soil in the entire belt. The problem is compounded further by mismanagement of soil or lack of appreciation for better water management strategy in the area. The impact is very visible today with increasing area under salinization/sodification, symptoms of multiple nutrient deficiencies and plateauing or declining yield in paddy and sugarcane.

3.8   PROBLEMS AND POTENTIAL AREAS IN THE HOBLI

The problem and potentials areas are identified from the database and maps showing their occurrence in each of the village were generated. As per this, soil depth is not a limiting factor for crop production in the area. Only in two soils the depth is less than 75 cm and not a constraint for most of the cultivated crops in the block. The dominant texture of the surface soil is clay, sandy clay, sandy clay loam, sandy loam and loamy sand. The surface soil texture in uplands is lighter (sandy loam, loamy sand) and heavy soil texture was found in lowlands and midlands, mostly clay and clay loam. The heavy texture results in poor aeration, impeded drainage, poor root penetration and waterlogging. Out of the eight soil series mapped in the uplands, Torebommanahalli (Tbl), Manigere (Mgr) and Yadaganahalli (Ydh) soil series have high amount of gravels (35–70%), which is a constraint for crop production.

In the Hobli, sheet and rill erosion are common in the upland and midland areas. Though, the slope is gently to very gently sloping in most of the areas, sheet and rill erosion is becoming a serious problem in the uplands due to the neglect. In the irrigated lands, due to terracing and leveling this problem is under control.

Soil drainage is a serious problem in about 1913.2 ha (11.3%) lands occurring mainly along Kadamba river and its tributaries. Moderately well drained soils occupy about 3171 ha (18.8%) area. Soil reaction has become a critical input for crop production in this Hobli, since it affects the availability of nutrients from soil. Many soil series identified in the block are towards moderate to strongly alkaline condition, particularly the lowland soils (Fig. 3.5). Even in the midlands and uplands, the soil reaction is more than 8.0 and in many soils, it is more than 8.5.

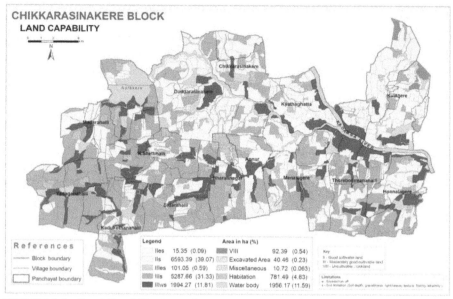

FIGURE 3.4   Land Capability Map of Chikkarasinekere Hobli.

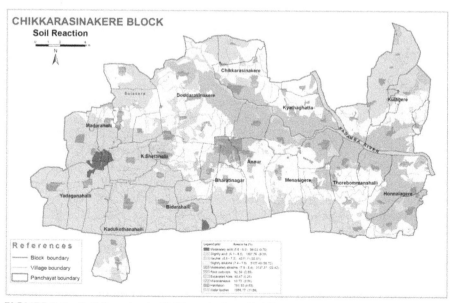

FIGURE 3.5   Soil Reaction in Chikkarasinekere Hobli.

3.8.1 LAND CAPABILITY ASSESSMENT

The land capability assessment (Klingebiel, A. A. and Montgomery, P. H 1961) showed an area of about 6609 ha area (39.2%) as good cultivable lands (Class II), 7383 ha (43.8%) as moderately good cultivable lands (class III) and 92.39 ha (0.55%) area as class VIII lands in the Hobli (Figure 3.4). Land irrigability classification showed an area of about 6609 ha (39.2%) as good irrigable lands and 7383 ha (44%) as moderately good irrigable lands (predominantly class III) These Class III lands have moderate limitation of heavy texture, depth, gravel and drainage.

3.8.2 SOIL SITE CROP SUITABILITY ASSESSMENT

The Land suitability assessment showed the suitability of the soil and other land resources at parcel level and the type of constraints that affect the yield and productivity of major crops grown in the block (FAO, 1976, 1983). As per this assessment, about 1879 ha (11%) area in the block is found to be highly suitable (S1) for rice cultivation, 5290 ha (31%) moderately suitable (S2) and 3178 ha (19%) marginally suitable (S3) for rice in the area. An area of about 3645 ha (22%) is found to be not suitable for rice. The major constraints found to be affecting paddy cultivation in Chikkarasinakere Hobli are gravelliness, topography, rooting depth, texture and drainage. The lowlands of Chikkarasinakere are generally found to be highly suited to rice.

For sugarcane, the assessment showed that soils of uplands and midlands are suitable while soils of lowlands are not those suitable for cultivation in Chikkarasinakere hobli. As per the assessment about 4053 ha (24%) are categorized as highly suitable (S1), 4855 ha (29%) under moderately suitable (S2) and 5088 ha (30%) under marginally suitable (S3) category for sugarcane production in Chikkarasinakere hobli (Fig. 3.6). The dominant limitations for successful production of sugarcane are gravelliness, salinity, calcareousness and rooting depth.

For ragi cultivation, subsurface salinity, sodicity and poor drainage conditions are the major constraints in this area. The suitability assessment for mulberry indicated that about 802 ha (4.8%) are highly suitable (S1), 3313 ha (20%) as moderately suitable (S2) and 9977 (59%) as marginally suitable. Heavy texture, inadequate drainage facility and waterlogged condition for more than four months pose very severe limitation in growing horticultural crops like mango and guava in the area. Similar assessments were carried out for all other crops cultivated in the area.

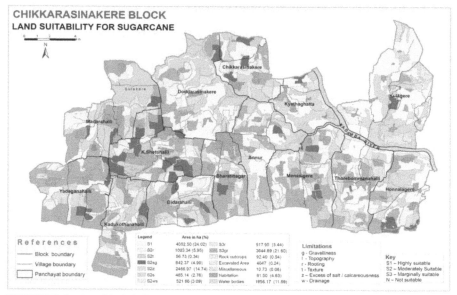

FIGURE 3.6   Land Suitability for Sugarcane in Chikkarsinkere Hobli.

3.8.3   IMPACT OF IRRIGATION ON DRAINAGE

The study has revealed that waterlogging is a major problem, particularly in all the lowlands, and to some extent in the upland areas of the block. The lowlands occupy about 4706 ha (29% of the area) in the Hobli. Normally the low lands are characterized by nearly level to flat topography and very deep, well drained, fine textured soils with good water holding capacity. They are the most productive soils of the Hobli. Even without any irrigation facility one assured crop under rainfed situation is possible and that was the scenario prevailed in most part of the Hobli before the introduction of K.R. Sagar canal irrigation during the 1905.

With the introduction of canal water and consequent change in the cropping pattern, from mostly rainfed crops to paddy and sugarcane, the microclimate has changed dramatically in the area over the years. The water released into the canals and their distributaries moved from the uplands to the midlands and finally into the lowlands. Due to its flat topography, the flow of the collected water is very much hampered. This has resulted in the rise of water table, slowly in the beginning and reaching the surface of the soil later (Photographs 3.3 and 3.6). The situation is compounded by the continuous addition of water from the upland and midland fields without any check or control. This has resulted in the stagnation of water for a longer period of time within the soil profile and at the surface.

Due to this, all the low lying areas, occurring particularly along the river and stream courses and tank command areas of the block are in a poorly to imperfectly drained condition. These soils occupy about 11% of the cultivable area in the block. Moderately well drained soils occur in about 19% of the cultivable area in the block, mostly in the transition areas above the lowlands.

The continuous of submergence of the soil under water has created significant changes in the physical, chemical and morphological properties of the soil in the lowland areas. The most visible change is in the soil color, soil structure, aeration, soil reaction, and in the microbial population of the soil (Photographs 3.6; Figure 3.5). The lack of oxygen has resulted in reduced condition, which in turn has affected the availability of few macro and micronutrients in the soil, thereby affecting the overall productivity of the soil. The poor drainage condition has reduced the option for crop selection and restricted the choice mainly to paddy and few other crops in the block.

Any area is considered water logged when either water stagnates on the land surface or the water table rises to an extent that soil pores in the crop root zone become saturated, resulting in restrictions in normal circulation of air, decline in the level of oxygen and increase in the level of carbon dioxide. In practical terms, a land is considered to be water logged when the water table is within +150 cm of the natural surface. The condition is also called "hypoxia" indicating that there is severe deficiency of oxygen in the root zone, which is essential for the normal growth of plants.

3.8.4   IMPACT OF WATER LOGGING

The primary effect of waterlogging on crop growth is from reduced soil aeration as a result of excess water. Reduced soil aeration around a crop's root zone results in decreased respiration, which reduces nutrient uptake, crop growth, and yield. In general, the performance of many field crops can be related to the depth of the water table. For most crops, there exists some water table depth, at which aeration, moisture, and nutrients are such that crop yields can be maximized. When the water table rises above this threshold, crop yields begin to decline The optimum water table depth will not only be a function of crop type, but will also be a function of other soil and climatic properties. Many studies have documented the negative effects of waterlogged soils on crop growth and yield. The extent of crop loss (Table 3.4) in the study area is very substantial.

3.8.5   WATER LOGGING AND ROOT ENVIRONMENT

All biological processes are strongly influenced by soil temperature. Water logged soils have large heat capacity and are relatively colder than dry soils. As a result, crop growth in waterlogged soils starts later and is slower than in dry soils. Since the compressive strength and the load bearing capacity of a soil decreases with increasing moisture content, it is relatively difficult to employ heavy machinery on waterlogged lands. A direct consequence of it is the delays in the sowing/harvesting operations of the crops and consequent yield reduction.

Bringing large quantities of water to perched lands without adequate provisions of drainage has resulted in an imbalance in the input and output of water causing in-equilibrium. The rising water table in such situations is a foregone conclusion. By acting as driver of change in land use, the massive irrigation systems have added another dimension to the twin problems of water logging and secondary salinization. Earlier studies have revealed that the small and marginal farmers are worst affected by soil degradation (Jai Singh and Jai Pal Singh, 1995). The large farmers have also experienced the brunt of soil degradation but the effect has been marginal since they have alternative sources of livelihood. The study has further indicated that the extent of inequity is higher on degraded than normal soils. However, this can be reduced to a great extent by launching land reclamation programs.

Development of salinity and its impact on productivity: When the drainage is poor, the dissolved salts move to the surface of the soil through capillary rise. With the prevailing high temperature and PET there is huge loss of water from the soil due to the higher rate of evapo-transpiration. This results in the accumulation of salts at the surface and subsequent development of salinity in the area. The impact of this is seen in almost all the lowland and midland soils identified in the block.

Out of the four series mapped in the lowland areas Chikkarasinakere and Doddarsinkere series have moderate to strongly alkaline condition and Madenahalli and Honnalagere have slightly to moderately alkaline conditions (Tables 3.1 and 3.2). The Kyathaghatta series occurring in the transitional areas between lowlands and uplands is also strongly alkaline in nature. Even in the uplands, many series have moderate to strong alkalinity problem. Only two series, namely Aravanahalli and Yadaganahalli, which occur in the uplands and not subjected to irrigation, have acidic to neutral reaction. Moderately alkaline soils occur in about 22% area and slightly alkaline soils occur in about 37% of the area in the block (Fig. 3.5).

Salinity affects crop growth by increasing the osmotic potential of the soil solution. In general, increased osmotic potential in the soil solution decreases a crop's ability to extract water and results in suppressed plant growth and decreased yield. Additional plant symptoms associated with high salinity levels are similar in appearance to those of drought, such as wilting and unusually green and thick leaves. The impact of salinity on crop yield in the block is very significant and based on the study, it was estimated that the yield loss in case of paddy cultivated on degraded areas when compared to normal areas is likely to be about 33% (Table 3.3). Similar loss can be expected in other crops also.

TABLE 3.1   Soil pH Ranges of the Soil Series Mapped in the Block

Soil Series	Ph (Range)	ESP (Surface soil)	EC (mSm−1)	Remarks
			Upland soils	
Kadakothanahalli	7.9–8.1	1.5	Traces	Moderately alkaline
Aravanahalli	6.0–7.1	0.5	Traces	Neutral
Yadaganahalli	5.6–7.3	0.3	Traces	Acidic to neutral
Kudagere	7.9–8.6	3.6	0.20	Moderate to strongly alkaline
Torebommanahalli	7.6–7.9	2.8	0.34	Slightly to moderately alkaline
Bidarahalli	7.7–8.0	0.85	0.09	Moderately alkaline
Manigere	8.3–8.7	5.8	0.37	Strongly alkaline
Honnanaya Kana-halli	7.1–7.9	4.5	0.25	Neutral to slightly alkaline
	Midland soils			
Kyathaghatta	8.9–9.4	6.0	Traces	Very strongly alkaline
	Lowland soils			
Chikkarsinkere	8.0–9.2	17	0.17	Moderate to Very strongly alkaline
Doddarsinkere	8.1–9.2	17	Traces	Moderate to Very strongly alkaline
Madenahalli	7.4–8.3	3.5	Traces	Slightly to moderately alkaline
Honnalagere	7.4–7.7	3.0	0.15	Slightly alkaline

TABLE 3.2   Surface Soil Reaction Classes

Reaction classes	Area in ha	Percent of the area
Medium to slightly acid	1406	10
Neutral (6.6–7.3)	4311	31
Mildly alkaline (7.4–7.8)	5138	37
Moderately alkaline (7.9–8.4)	3137	22

Though salinity has already become a serious problem in about 22% of the area in the block, it is emerging as a serious problem in another 37% of the soils mapped in the Hobli, particularly occurring in the transition areas lying between the uplands and lowlands. Apart from the general trend, in many villages salinity has become a major problem, seriously affecting the productivity. For example, in Kyathaghatta village, except Aravanahalli soil, which is not brought under irrigation, other soils have severe salinity/sodicity problem. Out of the 537 ha area in the village, more than 500 ha area has serious salinity problem. In this village, all lowland and mid-land soils, particularly Kyathaghatta, Chikkarasinakere and Doddarasinekere series have severe sodicity problem (Photograph 3.6).

The situation is similar in many other villages like Doddarsinkere, Chikkarsin-kere, Torebommanahalli, Hagalahalli, Madenahalli, Bannali, Menasigere and Mudenahalli villages in the block. Out of the 722 ha in Hagalahalli village, more than 600 ha area is having salinity problem. The situation is similar in almost all the villages located along the Shimsha river course and tank command areas of the block due to waterlogging and poor drainage conditions prevailing in the area. This can be clearly seen on the imagery of the area as light to dark gray tones. In such areas, even the growth and performance of paddy is very poor (Photograph 3.7). The leaves are thick and unusually green. It can be inferred with certainly that the productivity of other sensitive crops can be much less than what can be expected in such intensively cultivated areas of the block.

PHOTOGRAPH 3.7   Impact of Salinity on Paddy in Doddaarsinekere village.

3.8.6 SOIL NUTRIENT IMBALANCE

Due to intensive cultivation and monoculture system of cropping (paddy/sugarcane) followed in the region for more than 80 years, a severe plant nutrient imbalance is noticed in large parts of the block with respect to all the nutrients studied. Under such circumstances the management of nutrients becomes very complex as there are complex nutrient interactions involved. This itself becomes a major constraint in realizing the desired yield levels. This calls for intensive monitoring of nutrient status after every cropping season and undertaking corrective steps. Best option is to adopt integrated nutrient management strategy involving, organic, chemical and biological nutrient sources and regular practice of crop rotation (Table 3.3).

The available potassium content has become low in 38% of the area in the Hobli. It is predominantly low in in Chikkarasinakere, Kyathaghatta, Yadaganahalli, Doddarasinakere, Kadukothanahalli and Bidarahalli Panchayats and medium in Annur, Menasigere, Torebommanahalli, Hagalahalli, Kulagere, Bharathinagar and K. Shettihalli panchayats.

TABLE 3.3 Nutrient Imbalance in the Block (%area)

Available Nutrient	Deficient	Excess	Sufficient
Nitrogen	20	50	30
Phosphorus	10	38	52
Potassium	38	2	60
Zinc	30	-	70

The available phosphorus status is high in 38% of the Hobli. It is predominantly high in Bidarahalli, Bharatinagar, Menasigere and Torebommanahalli panchayats. However, parts of Yadaganahalli, Madarahalli and K. Shettihalli panchayats (7% area), the available phosphorus status is low. The available zinc content is deficient in Kadukothanahalli, Bidarahalli, K. Shettihalli, Bharathinagar, Kulagere and Torebommanahalli panchayats and sufficient in Menasigere and Hagalahalli panchayats.

3.8.6 APPROXIMATE LOSS TO THE LOCAL ECONOMY DUE TO THE DEVELOPMENT OF SALINITY/SODICITY AND WATERLOGGING (TABLE 3.4)

If one ha of irrigated land goes out of cultivation, then more than rupees one lakh/ha, invested in the development of irrigation potential in the area is blocked. The interest on this investment comes to Rs. 10,000 per annum. As indicated in the Table 3.4, the extent of crop loss for paddy in the block per year is about Rs 37.5 crores. The cost of cultivation gets escalated at least by Rs 5000/ha/season due to nutrient imbalance, waterlogging and salinity/sodicity. This amounts to Rs 10 crores per year for the entire block. As per the survey, about 10,030 ha area in the block is facing the problem of degradation. Accordingly, the total loss in the block per year is about Rs 47.5 crores.

If we assume that the situation is more are less similar in the entire Cauvery command area (2.8 Lakh ha), as the terrain and topography are repeating in the region, then the value of crop loss is about Rs 630 crores and when we add loss due to increased cost of cultivation, the total loss per year is about 795 crores. This excludes the losses to investment/interest on investment made in the creation of irrigation infrastructure by the government. If the interest on investments on irrigation infrastructure is also taken into account, nearly about Rs 1000 crores worth economic loss is taking place due to the development of salinity/sodicity, water logging and multiple nutrient deficiencies in the command area. In the coming years, the situation is likely to worsen further and the expected loss is likely to be higher than what is estimated at present in the command area.

TABLE 3.4 Economic Consequences of Land Degradation in the Cauvery Command Area—A Case of Paddy Production System

No	Items	Details	Remarks
1	Farming area in the Chikkarasinakere block	17,000 ha	
2	Extent of area facing land degradation in the block (59%)	10,030 ha	Salinity, alkalinity
3	Paddy productivity under recommended level of package of practices in the area	5.5t/ha	UAS(B) PP
4	Normal productivity of paddy under farmers practices	4.5t/h	
5	Paddy productivity in degraded lands	3.0t/ha	Data from farmers
6	Paddy yield loss in the block due to land degradation	30,000 tons	2 crops/yr
7	Value of paddy yield loss in the block per year	Rs 37.5 crores/ year	Support price Rs 1250/q
8	Losses due to increased cost of cultivation under imbalanced nutrient status in the block	Rs 10 crores	@Rs 5000/- per season
8	Area irrigated under Cauvery command in Karnataka	280,000 ha	
9	Area likely to be affected with degradation process	165,200 ha	59% area
10	Paddy yield loss (@33%)	840,000 tons	2 crops/year

TABLE 3.4   *(Continued)*

No	Items	Details	Remarks
11	Value of crop loss	Rs 630 crores	Support price Rs 1250/q
12	Losses due to increased cost of cultivation under imbalanced nutrient status in the command are	Rs 165 crores	
13	Total loss per year in the entire command area	Rs 795 crores	
14	Approximate loss of interest on the investments made on irrigation infrastructure	Rs 200 crores	
15	Approximate total loss	Rs 995 crores	

3.9   STATE OF LAND RESOURCES AND ITS MANAGEMENT NEEDS

The farm-level survey has brought out clearly the site-specific status of soil, water and other resource base in the hobli, their constraints, potentials and their suitability along with their management needs. The development of salinity, particularly in the canal irrigated areas of the block, was not brought out by the earlier small-scale surveys carried out in the area and also in the district. This cadastral level survey has not only identified the survey number wise problems and their suitability for various uses but also the interventions required to overcome the same at the field level. For example, the farm level survey has revealed that waterlogging is a major problem, particularly in all the lowlands, and to some extent in the upland areas of the block, which is responsible for the development salinity in the area.

Though salinity has already become a serious problem in about 22% of the area in the block, it is emerging as a serious problem in another 37% of the soils mapped in the hobli, particularly occurring in the transition areas lying between the uplands and lowlands. In many villages salinity has become a major problem. For example, in Kyathaghatta village, except Aravanahalli soil, which is not brought under irrigation, other soils have severe salinity/sodicity problem. Out of the 537 ha area in the village, more than 500 ha area is affected by salinity. The situation is similar in many other villages in the block. This can be clearly seen on the imagery of the area and also at the field level (Photograph 3.7).

3.10   RELEVANCE OF THE STUDY IN THE NATIONAL CONTEXT

Land and water are the two most important natural resources for agricultural development and economic advancement of any country. With a low per capita availability of land and water in India compared to other countries, enhancing agricultural production has become very essential to meet the growing food demands of the population. Thus, available water for irrigation needs to be used judiciously. At the same time, land degradation due to soil salinity and water logging is threatening the sustainable use of these resources. Efforts to address salinity and waterlogging problems as they arise will prevent substantial losses to farmers and to the regional economy. This has also great social and environmental advantages. One should also take into account that in arid and semiarid areas, irrigation systems require a proper drainage system. Without a drainage system such irrigation systems are unsustainable. The feasibility of the drainage systems in such areas should, therefore, always be assessed in combination with the irrigation systems. The case study clearly demonstrates the utility of farm wise land resources database in understanding the constraints and potentials and aids in devising the most appropriate management plans.

3.11 CONCLUSION

The present database will help in the holistic treatment and management of the problems identified in each and every village, particularly the fast emerging issue of salinity in the Hobli. Further it can form the basis for formulating various management strategies required for increasing productivity and conservation of the resource base. This can be used for promoting precision farming at local level and will be helpful in equipping the farmers of the Hobli with required information so that they are able to meet the challenges of the future in an effective manner.

Each management unit is unique in the properties and requires specific conservation plan and management strategies to improve the plot wise productivity. Database is presented in a GIS environment to facilitate easy reference and retrieval. Various interpretative theme maps like soil depth, texture, drainage, slope, Available Water Capacity, Soil organic carbon status, Land capability, Prime farmlands, soil fertility (NPK and micronutrients), soil-site suitability for major crops are presented for ready use by the developmental departments. Information provided in report and maps can also be easily referred by the stakeholders to identify field survey number with best land use options.

At present the Government of India has not made it obligatory for the generation of farm wise detailed land resources database for the agencies executing the watershed development programs. There is a need to give legal status/backup by Government of India, so that all the land based rural development programs are based on the farm specific land resources database and are truly science based.

KEYWORDS

- **Alkalinity**
- **Aravanahalli**
- **Cadaral Sheets**
- **Canal Irrigated Lands**
- **Chikkarasinkere**
- **Doddarsinkere**
- **Evapo-transpiration**
- **Food Security Bill**
- **Green Revolution**
- **K.R. Sagar canal irrigation**
- **Kyathaghatta**
- **Lowlands**
- **Remote Sensing Data Product**
- **Salinity**
- **Uplands**
- **Valleys**
- **Wastelands**
- **Waterlogging**
- **Yadaganahalli**

REFERENCES

FAO (1976). A Framework for Land Evaluation. FAO Soils Bulletin No. 32, FAO, Rome.

FAO (1983. Guidelines: Land Evaluation for Rainfed Agriculture. FAO, Rome.

Jackson, M. L. (1973). Soil Chemical Analysis. Prentice Hall of India Pvt. Ltd. New Delhi.

Jai Singh and Jai Pal Singh 1995. Land degradation and economic sustainability, Ecological Economics. 15A (1), 77–86.

Klingebiel, A. A., Montgomery, P. H. (1961). Land capability Classification, USDA Handbook 210, Washington, DC.

Natarajan, A., Dipak Sarkar (2010). Field Guide for Soil Survey, NBSS&LUP, Nagpur, Pages 71.

Soil Survey Staff (1951). Soil Survey Manual, Agriculture Handbook No.18, Washington, DC.

Soil Survey Staff (1993). Soil Survey Manual, Agriculture Handbook No.18, USDA, Washington, DC.

Soil Survey Staff (1996). National Soil Survey Handbook, Title 430-VI, NRCS, USDA, Washington, DC.

CHAPTER 4

DELINEATION OF PRIME AGRICULTURAL LANDS FOR LAND USE PLANNING—A CASE STUDY OF MYSORE DISTRICT, KARNATAKA

V. RAMAMURTHY[1], K. M. NAIR, L. G. K. NAIDU, and DIPAK SARKAR

CONTENTS

National Bureau of Soil Survey and Land Use Planning, Regional Centre, Bangalore-24.

ABSTRACT

Agricultural lands are under pressure in India as in many other countries for competing other land uses. This chapter presents concept and criteria's and methodology for delineating prime and marginal lands by spatial integration of external land features, soils, agro-ecology, present land use and farming systems. Mysore district is having 26.9% irrigated prime lands, 43.5% rainfed prime lands and 4.5% marginal lands. Both irrigated and rainfed prime lands needs to be preserved to meet the growing food demand and marginal lands can be diverted to other nonagricultural uses like industrial, infrastructural development, etc.

4.1   INTRODUCTION

Planning and management of land resources are integral part of any developmental program. The importance of land as a resource cannot be overemphasized and one of the major global concerns is the problem of declining agricultural productivity and land resources that are being threatened by the rapid human population growth. Sustainable land and natural resource management is a key factor in ensuring adequate food, fodder, and fiber production. Land use intensification due to demographic pressure led to expansion of agricultural land in to fragile systems and it is one of the major causes of degradation of natural resources. Despite the impressive gains in agricultural production and productivity through crop intensification, most of the regions are facing problems related to the land degradation due to intensive cultivation, the overuse of ground water, excessive nutrient loads in surface and ground water, and increased pesticide use. A sustainable and increasingly productive agricultural base is essential for food security. One aspect of land reform that has not been extensively addressed is its relationship to household decision making regarding patterns of land use and agricultural practices. Considering the multiple demands of land and water resources, the time has now come to take hard decisions in planning limited natural resources right from the lowest possible level to fulfill the aspiration of all the stakeholders. At the local level, the stakeholders' knowledge of sustainable land management practices relies on time-tested traditional technologies that have been transferred through generations. In some cases, what might have been a sustainable land use practice in the past may not be viable anymore in the new context. There is an increasing need to use resources in a sustained manner to increase the production but at the same time protecting the environment, biodiversity and global climate systems. The classification and mapping of agricultural lands according to the capability and utility are important prerequisites for better management. It provides a scientific basis for resource management, conflict resolution and decision-making at the local and regional levels, consistent with national policies and priorities. However, the existing database on land use is highly inadequate. Therefore, strengthening of the database, using traditional cadastral surveys, modern remote sensing techniques, GPS, GIS and computerization of land records would be necessary.

Issues related to land use are becoming increasingly complex and land use planners/managers/officials often lack the right information at their disposal, thus lowering their capacity to make judicious land use options/land management decisions. Sectored and single objective approaches to land use planning, advocated in the past are no longer are relevant and effective. **The necessity of developing comprehensive land use plans at different levels has been increasingly felt and emphasized in different five-year plans. In this context, identification and delineation of prime agricultural lands helps land use managers to take appropriate decisions for sustainable agricultural production and food security besides conserving precious natural resources.**

Conversion of agricultural land to nonagricultural uses is more intense in recent years in India. The causes for loss of agricultural lands are almost self-evident. They include, profits from land sales for nonfarming purposes, conversion of agricultural lands for housing, and development of highways, industries, construction of airports, etc. Prime agricultural lands need to be preserved for future food security and prosperity of rural areas. Delineation of prime agricultural lands at different levels helps in spatial zoning policy (Special Agricultural Zones), where zoning of land use ensures and stimulates optimal use of the land on a permanent basis. Such preservation strategies require a vision of how agriculture should develop and which farmland should be preserved in order to achieve this vision.

Prime farm (agricultural) land, as a designation assigned by U.S. Department of Agriculture, is land that has the best combination of physical and chemical characteristics for producing food, feed, forage, fiber, and oilseed crops and is also available for these uses. It has the soil quality, growing season, and moisture supply needed to produce economically sustained high yields of crops when treated and managed according to acceptable farming methods, including water management. In general, prime agricultural lands have an adequate and dependable water supply from precipitation or irrigation, a favorable temperature and growing season, acceptable acidity or alkalinity, acceptable salt and sodium content, and few or no rocks. They are permeable to water and air. Prime farmlands are not excessively erodable or saturated with water for a long period of time, and they either do not flood frequently or are protected from flooding (Soil Survey Staff, 1993). The framework of delineating prime agricultural lands proposed by USDA was followed with necessary modification to suit local conditions of Mysore district. Indicators proposed are specific to Mysore district to identify prime lands from available land area and indicators may not be same in other parts of the country. According to the framework of USDA, one may or may not get prime lands at all in our situation.

The objective of this chapter is to interpret the soil survey information available at 1:50,000 to identify and delineate prime agricultural lands at district level for optimum utilization of natural resources and to help land use planners/managers/

policy makers to take appropriate decisions to preserve precious land for future generations.

4.2 METHODOLOGY

The general methodology followed in identification and delineation of prime agricultural lands in Mysore district is explained (Fig. 4.1) in this section. Soil resource map (1:50,000 scale) of Mysore district was generalized considering landform, depth, texture, erosion and gravellines. Land use/land cover map obtained from Karnataka State Remote Sensing Application Centre was generalized into 7 from 23 classes. In order to generate agro-ecological units (AEU) map of the district 25 years rainfall data from 44 rain-gauge stations were analyzed for probability of occurrence of dry spells in premonsoon, post monsoon and calculated Length of Growing Period (LGP). AEUs were delineated with Hobli (Block) as the lowest administrative unit. Indicators used in delineation of prime irrigated, rainfed and marginal soils are presented in Table 4.1. Major production systems of the district in irrigated and rainfed situations are identified. Integrated generalized soil, land use/land cover, AEU, and production systems in GIS environment to identify homogenous land units called Land Management Units (LMU). Overlaid village boundaries map on the LMUs to generate village-wise area statistics. LMUs were further grouped into prime irrigated, prime rainfed and marginal agricultural lands at district level.

TABLE 4.1   Indicators Used in Delineation of Prime Lands

Particulars	Prime irrigated	Prime rainfed	Marginal lands
	Indicators		
Soil depth	-	>50 cm	<50 cm
Landform	-	Level to gently sloping	Moderate to Highly sloping lands
Gravelliness	Non gravelly	Non gravelly to slightly gravel	Gravelly
Length of growing period	-	>120 days	<120 days
Erosion	-	Nil to slight	Moderate to severe
Productivity of crops	<30% deviation from attainable yield	<30% deviation from attainable yield	>30% deviation from attainable yield

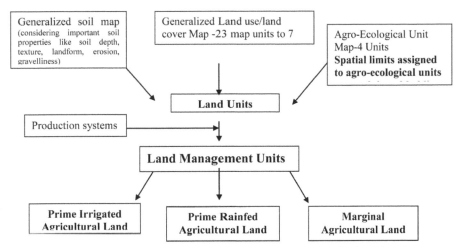

FIGURE 4.1   Methodology Followed for Delineation of Prime Agricultural Land.

4.3   RESULTS AND DISCUSSION

Land units were delineated by spatial integration of external land features, soils, agro-ecology, present land use and administrative divisions and identified 96 land units.

Farming systems data including productivity of different crops was collected from all the major land units. In Mysore district, five major production systems are prevailing in irrigated situation two system, that is, rice and sugarcane production with dairy and rice and maize production system with Dairy+Sericulture. Whereas in rainfed conditions, Tobacco-ragi/pulses, Cereal (Ragi, Jowar, Maize), and Cotton based production systems. In rainfed situation, integrated mixed farming system is dominant (2–3 cows/buffalo + pair of bullocks, 8–10 sheep/goats).

Land units were further grouped into homogenous LMUs based on similar farming system/cropping systems existing in the different land units. 12 LMUs were delineated in Mysore district. LMUs are further grouped into prime irrigated lands, prime rainfed lands (Fig. 4.2) and marginal lands as per the indicators given in Table 4.1.

All irrigated LMUs (LMU 1, 2 and 3) were grouped in to irrigated prime lands because these LMUs are having more than 30% area under irrigation, rainfed LMUs (LMU 5, LMU 6, LMU 7, LMU 9, LMU 11, LMU 12) having medium deep (50–100 cm) to deep (100–150 cm) soil with crop growing period (LGP) of 120 days and above, slight to moderate erosion and yield gap of crops less than 30% were grouped into prime rainfed lands. LMUs having shallow soils with less than 120 days LGP and moderately to severely eroded soils were grouped in to marginal lands (LMU 4, LMU 8, LMU 10).

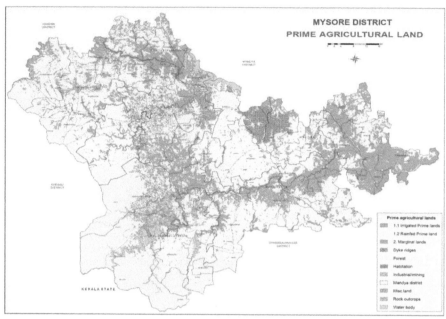

FIGURE 4.2   Prime Agricultural Lands in Mysore District.

Distribution and percent area under prime irrigated and rainfed and marginal lands under different taluk wise presented in Table 4.2. T.Narsipura taluk is having highest prime irrigated lands followed by K.R. Nagar and Hunsur. Minimal prime irrigated lands are in Piriyapatna taluk. Nanjangud taluk is having highest prime rainfed area followed by Piriyapatna, Mysore, H.D. Kote and Hunsur taluks. However, most of the prime rainfed lands of Mysore and Nanjanagud taluks have been included in the recent Mysore Urban Development Authority (MUDA) plan. There is a need for Government agencies to consider the potentials of land before earmarking for other than agricultural uses. Hunsur and Piriyapatna taluks have comparatively more marginal lands. Marginal lands can be diverted to other than agricultural uses.

TABLE 4.2   Area Under Different Category of Prime Lands in Mysore District

Taluks	Percent area		
	Prime Irrigated Lands	**Prime Rainfed Lands**	**Marginal Lands**
H.D. Kote	4.25	6.91	0.63
Hunsur	4.50	6.46	0.95
K.R. Nagar	4.81	3.50	0.14
Mysore	1.27	7.70	0.68

TABLE 4.2   *(Continued)*

Taluks	Percent area		
	Prime Irrigated Lands	**Prime Rainfed Lands**	**Marginal Lands**
Nanjanagud	.24	8.71	0.74
Piriyapatna	2.08	7.77	0.91
T. Narsipur	5.72	2.47	0.41

Similarly distribution of different category of prime land in different hobli wise is presented in Table 4.3. H.D. Kote, Hunsur and Nanjangud hoblis are having highest prime irrigated lands compared to other hoblis. Jayapura hobli of Mysore, Hanagodu hobli of Hunsur and Hullahalli hobli of Nanjangud are having highest prime rainfed lands compared to other hoblis. Whereas, Bilikere hobli of Hunsur and Hampapur hobli of H.D. Kote are having highest area under marginal lands.

TABLE 4.3   Area Under Different Category of Prime Lands in Different Taluks of Mysore District

Taluk	Hobli/Block	Percent area		
		Prime irrigated Lands	**Prime Rainfed Lands**	**Marginal Lands**
H.D. Kote	Anatharasanthe	0.68	0.85	0.00
	Hampapura	0.81	1.24	0.63
	H.D. Kote	1.92	1.64	0.00
	Kandalike	0.80	1.43	0.00
	Saragur	0.76	1.75	0.00
Hunsur	Bilikere	0.81	1.04	0.88
	Gowdagere	1.19	1.05	0.01
	Hanagodu	0.74	3.13	0.00
	Hunsur	1.76	1.24	0.06
Mysore	Elawala	0.11	1.74	0.14
	Jayapura	0.19	3.30	0.08
	Mysore	0.37	0.70	0.11
	Varuna	0.60	1.95	0.35
Nanjanagud	Biligeri	0.90	0.78	0.06
	Chikkannachatra	0.58	1.29	0.15
	Hullahalli	1.19	2.66	0.01
	Kowlande	0.14	2.45	0.26
	Nanjanagud	1.43	1.53	0.26

In prime irrigated lands, most economical and suitable crops are Banana, Sesame/Cowpea/Black gram/Green gram/Green manure-rice, Rice-rice, Rice, Sugarcane, Ragi, Vegetables, Mulberry, Maize, Napier bajra hybrids on bunds with cow/Buffaloes and inland fishery in low laying/salt affected areas. Tobacco ragi/Horse gram/Field bean/Cowpea, Maize/cowpea-ragi, Ragi+Field bean/Red gram, Maize, Groundnut, Cotton, Jowar are the most suitable crops in prime rainfed areas of the district. Animal husbandry activities like rearing of 2–3 cows/buffaloes, sheep and goats and poultry farms provide livelihood security in prime rainfed areas. The marginal agricultural lands are mostly suitable for nonagricultural purposes (industries/Urban). However, growing of millets, pulses, improved forage varieties and agri-horti-silvi pasture helps to sustain the existing human and livestock population besides conserving natural resources.

4.4   CONCLUSION

Identification of different category of lands like prime irrigated and rainfed and marginal lands is the need of the hour to meet the growing population. Prime lands need to be preserved for meeting the food security of nation and marginal lands which are not profitable for farming can be allocated for other than agricultural uses. In Mysore district, study showed that 26.9 and 43.5% prime irrigated and rainfed lands are available and at the same time prime rainfed lands more prone to urbanization there by affecting the food security of the district.

KEYWORDS

- **Agri-Horti-Silvi Pasture**
- **Agro-Ecological Units**
- **Land Management Units**
- **Length of Growing Period**
- **Mysore Urban Development Authority Plan**

REFERENCES

Soil Survey Staff (1993). "Soil Survey Manual." Soil Conservation Service. U. S. Department of Agriculture Handbook 18.

USDA Bulletin No.1. Planning and Development Information Series.

CHAPTER 5

SOIL AND CROP HISTORY IN DOMINANT AGRO-ECOSYSTEMS OF THE INDO-GANGETIC PLAINS, INDIA

T. BHATTACHARYYA[1], P. CHANDRAN[2], S. K. RAY[4], P. TIWARY[3], C. MANDAL[4], DIPAK SARKAR[5], AND D. K. PAL[6]

CONTENTS

[1]Principal Scientist and Head, Division of Soil Resource Studies, National Bureau of Soil Survey and Land Use Planning (ICAR), Amravati Road, Nagpur 440 033, Maharashtra, India

[2]Principal Scientist, Division of Soil Resource Studies, National Bureau of Soil Survey and Land Use Planning (ICAR), Amravati Road, Nagpur 440 033, Maharashtra, India

[3]Senior Scientist, Division of Soil Resource Studies, National Bureau of Soil Survey and Land Use Planning (ICAR), Amravati Road, Nagpur 440 033, Maharashtra, India

[4]Principal Scientist and Incharge, Cartography Section, National Bureau of Soil Survey and Land Use Planning (ICAR), Amravati Road, Nagpur 440 033, Maharashtra, India

[5]Director, National Bureau of Soil Survey and Land Use Planning (ICAR), Amravati Road, Nagpur 440 033, Maharashtra, India

[6]Former Principal Scientist and Head, Division of Soil Resource Studies, NBSS&LUP, Nagpur and now Consultant, ICRISAT. Kamal Narayan Apts Q 3 Laxmi Nagar Nagpur-440022

ABSTRACT

The present study deals with detailed soil and crop information in various agro-ecosystems of the Indo-Gangetic Plains, India. The database generated through the Soil and Terrain Digital Database (SOTER), GEFSOC modeling system, soil survey and agro-ecological subregions indicate that role of soils is extremely important in maintaining ecosystem and climate regulation The study stressed on quantitative database soil information to generate a Soil Information System for the Indo-Gangetic Plains (SISIGP) and to dovetail the crop history to help the modelers, planners and land resource managers.

5.1   BACKGROUND

It has long been felt that our natural soil environment should be mapped and monitored with active participation of agencies responsible for managing natural resources, industry groups and community organizations. This organized information forms a basis for storing soil and land database for implementation and monitoring various efforts on land resource management. In view of huge demands on natural resources like soil and water, with special reference to environment and its protection there is a need for better information on spatial variation of soils and landscapes. It would not only store the datasets for posterity but also will improve our understanding of bio-physical processes in terms of cause-effect relationship in the pedo-environment to understand soil health. Thus, information on soil and land resources is fundamental and therefore soil information system (SIS) plays a pivotal role. Information on soil is huge but scattered and therefore it requires to be archived. Now since modern day information system of any natural resources requires its physical location in terms of space, exact referencing of important spots has become very necessary. Geographic Information System (GIS) has been an important tool for Geo-Referencing Soil Information System (GeoSIS). Since India is a large country we have decided to restrict this article to SISIGP for brevity (Fig. 5.1).

IGP covers 13% area of the country produces 50% of food grains and feed 40% of Indian population. Land use pattern of IGP is shown in Fig. 5.2. The major concerns of IGP have been mentioned by Abrol and Gupta (1998), Bhandari et al. (2002), and Gupta (2003). Reports on the recent land use and soils of the IGP indicate a general decline in soil fertility (Bhandari et al., 2002; Gupta, 2003). Soils that previously had a high nutrient status are now found to be deficient in many nutrients. Long-term fertility studies have shown reduction in organic matter and other essential nutrients (Abrol and Gupta, 1998; Bhandari et al., 2002). This indicates that biological activity of soils in maintaining the inherent soil fertility has been reduced over the years of cultivation. Thus, the soils need use of chemical fertilizers in maintaining their fertility (Abrol and Gupta, 1998). IGP is blessed with the perennial supply of water through the river Ganges, which is 2506 km long to form a valley or a basin (322–644 km wide). The river flows eastwards and empties into the Bay of Bengal. The combined delta is the largest in the world and is largely a tangled swampland. IGP represents the states of Punjab, Haryana, Delhi, Himachal Pradesh, Uttar Pradesh, Bihar, West Bengal and parts of Rajasthan and Tripura with 14 agro-ecological subregions and cover 52.0 mha area (Mandal et al., 2013). The details of the climate and other features are shown in Figs. 5.3–5.5.

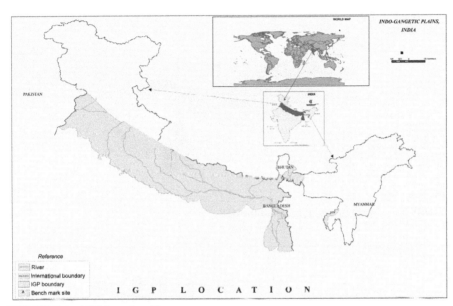

FIGURE 5.1 IGP Location Map.

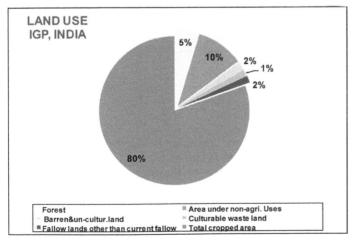

FIGURE 5.2 Land Use in IGP.

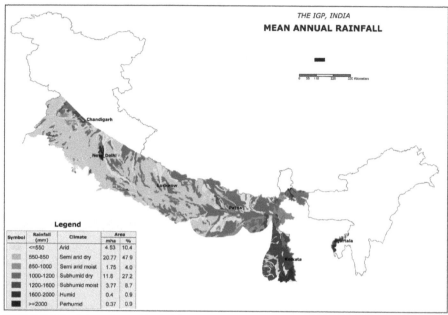

FIGURE 5.3   Mean Annual Rainfall of the IGP.

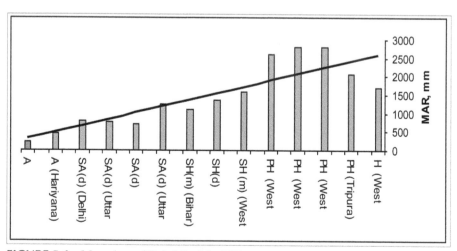

FIGURE 5.4   Mean Annual Rainfall of the IGP.
(H= humid; PH = Perhumid; SHm = Sub-humid, moist; SHd-Sub-humid, dry; Sad =. Semi-arid, dry; A = arid).

Soil information system was developed using SOTER technique which is soil and terrain analysis depicting a distinctive, often repetitive pattern of land form, lithology, surface form, slope, parent material and soils. SOTER technique was used in the IGP (SOTER-IGP), which allows mapping and characterization of land using the available information on soils as shown in Figs. 5.6 and 5.7. SIS-IGP permits linkages between soil profile data and the spatial component of a SOTER map for environmental applications, which in turn requires generalization of measured soil (profile) data by soil unit and depth zone. The set of soil parameter estimates for IGP should be seen as the best estimates based on the currently available selection of profile data held in IGP-SOTER and WISE (Batjes et al., 2007). Recently NAIP project on "Georeferenced Soil Information System for Land Use Planning and Monitoring Soil and Land Quality for Agriculture" has broadened datasets from the existing level (Chandran et al., 2012). The primary and secondary datasets for IGP are useful for agro-ecological zoning, land evaluation and mapping of carbon stocks and changes. Available soil and land information system in the IGP has been shown on various levels in Tables 5.1–5.13.

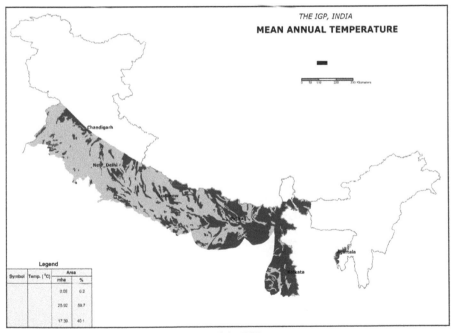

FIGURE 5.5   Mean Annual Temperature of the IGP.

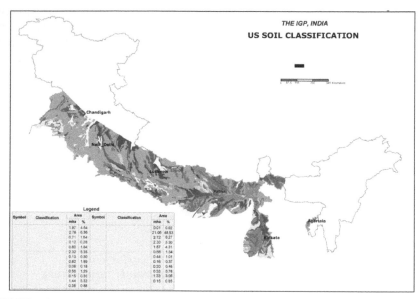

FIGURE 5.6   Soils of the Indo-Gangetic Plains (U.S. Soil Taxonomy).

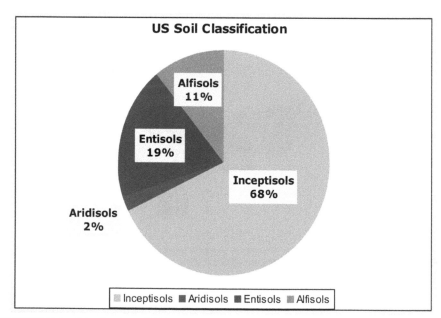

FIGURE 5.7   Distribution of Different Soil Orders Following US Soil Classification in the Indo-Gangetic Plains.

To develop user-friendly datasets, soil information was collected through different sources and at various levels (Tables 5.1 and 5.2). The Indo-Gangetic Plains (IGP) were mapped using soil mapping units according to the scale of mapping using the standard methods of soil survey. Level 1 on soil information system distinguishes major physiography, agro-ecological regions, agro-ecological subregions in the IGP. It provides information on climate, such as temperature and rainfall and a few selected physical, chemical and mineralogical properties of soils (Table 5.3). Soil information for developing this system has been collected from various sources to prepare mapping units of bio-climatic systems, agro-ecological regions, agro-ecological subregions and agro-climatic zones (Tables 5.4–5.7). The climate of the landscape also helps estimating the length of growing period in each region to select a crop for exact planning. Table 8 indicates various descriptions of special information in the IGP at district level in 1:50,000 scale. It details the physiographic unit at a final level to show slope, relief, vegetations, soil orders and then soil series level information. Table 5.9 shows the available datasets of soil information system at the watershed level (1:2400 scale). Tables 5.10 – 5.13 details soil information in various parts of the IGP covering the states of Rajasthan, Punjab, Haryana, Delhi, Uttar Pradesh, Bihar, West Bengal and Tripura.

TABLE 5.1   Available Soil and Land Information System: Spatial Hierarchy in the IGP (Level 1)

Level	Land Unit	Soil Unit	Descriptive Legends	Description of map unit	Map Scale	Source/Comments
1.	Country	Order	Soil Orders	Inceptisol, Entisols	1:25 million	NRCS (1996)
2.	State	Suborder	Soil Suborder		1:7 million	NBSSLUP (1985)/Map printed by NBSS and LUP, Nagpur
3.	State	Old soil classification	Traditional soil names	Red and yellow soils, Red Loamy soils, Mixed Red and Black soils	1:4 million	Govind Rajan (1971)
4.	State (Region)	–	Agro-ecological region (AER)	Bengal Plains, hot subhumid to humid LGP 210–300 days (AER 15)	1:4.4 million	Sehgal et al. (1992)/Map printed by NBSS and LUP, Nagpur

TABLE 5.1   *(Continued)*

Level	Land Unit	Soil Unit	Descriptive Legends	Description of map unit	Map Scale	Source/Comments
5.	State (Sub-region)	–	Agro-ecological subregion (AESR)	Bengal Basin and north Bihar Plains,, hot moist subhumid with medium to high AWC and LGP (210–300 days) (AER 15.1)	1:4.4 million	Velayutham et al. (1999)/ Map printed by NBSS and LUP, Nagpur
6.	Country	Soil family	Soil family association	Total 1649 units in the country – IGP had 74 no. of units.	1:1 million	NBSS and LUP (2002)/ (printed by NBSS and LUP)

TABLE 5.2   Available Soil and Land Information System: Spatial Hierarchy in the IGP (Level 2)

Level	Land Unit	Soil Unit	Descriptive Legends	Description of map unit	Map Scale	Source/Comments
7.	State (Physiography)	Soil family	Soil family association	Total 74 soil map units showing association of dominant (60% average in polygon) and subdominant (40% in a polygon soils); and 50%, 30% and 20% where 3 soil families exist.	1:250,000	Map generated during GEF-SOC Project for IGP, India (Bhattacharyya et al., 2005a)
8.	District	Soil Series	Soil series association	Total 28 soil units showing association of dominant and subdominant soil series with inclusions	1:50,000	NBSS and LUP (Map of Hooghly district) (Sarkar et al., 2001)

TABLE 5.2 *(Continued)*

Level	Land Unit	Soil Unit	Descriptive Legends	Description of map unit	Map Scale	Source/Comments
9.	Watershed	Soil series	Soil series association	Two soil series and five map units in Chuchura farm	1:2400	NBSS and LUP (Report No 448 (NBSS and LUP 1983)

TABLE 5.3 Description of Spatial Information in IGP (Level 1)

Particulars	Example	Remarks
Physiography	Alluvial Plains	SISIGP displays such information in selected AESR of India
Climate	Mean annual temperature, mean annual rainfall, rainy days, cloudiness, relative humidity, PET, vapor pressure, wind velocity	SISIGP gives such information
Soil	Physical, chemical and mineralogical properties of soils	Selected soil properties are available at this level

TABLE 5.4 Description of Spatial Information in Different Bioclimatic Systems of the IGP (Level 1)

Descriptions	Soil Series
Arid (hot): MAR <550 mm	Masitawali, Nihalkhera, Hisar
Semi-arid (hot): MAR 550–950 mm	Phaguwala, Ghabdan, Bhanra, Patiala, Zarifa Viran, Sonepat, Holambi, Sakit, Hirapur, Itwa, Bara
Sub-humid (dry): MAR 950–1100 mm	Dhadde, Jagjitpur, Fatehpur, Khiranwali, Berpura, Haldi, Bijapur, Basiaram, Simri, Sarthua
Sub-humid (moist): MAR 1100–1500 mm	Bahraich, Gaupur, Belsar, Nanpur, Ekchari
Sub-humid to humid: 1500–1800 mm	Madhpur, Hanrgram, Sasanga, Konarpara, Amarpur, Bansghatta, Seoraghuri
Per-humid: >1800 mm	Singhvita, Nayanpur, Khowai
Coastal: MAR varies	Sagar

Bhattacharjee et al. (1982); Ray et al. (2005).

TABLE 5.5 Description of Spatial Information in Different Agro-ecological Regions (AERs) of IGP (Level 1)

Descriptions	Soil Series
Western Plains, hot arid ecoregion, LGP: <60 days (AER 2)	Masitawali, Nihalkhera, Jassi Pauwali, Jodhpur Ramana
Northern Plains, hot semi-arid ecoregion, LGP 90–150 days (AER 4)	Fatehpur, Phaguwala, Zarifa Viran, Ghabdan, Bijapur, Hirapur, Sakit
Northern Plains, hot sub-humid (dry) ecoregion, LGP 120–180 days (AER 9)	Dhoda, Jagjitpur, Bhanra, Berpura, Basiaram, Itwa, Simri, Akbarpur
Eastern Plains, hot sub-humid (moist) ecoregion, LGP 180–210 days (AER 13)	Bahraich, Haldi
Bengal Plains, hot sub-humid to humid (inclusions of perhumid ecoregion), LGP 210–300 days (AER 15)	Sasanga, Konarpara, Hanrgram, Amarpur, Madhpur, Barak, Seoraguri
Eastern Himalayas, warm per-humdi AER, LGP 270–300 days (AER 16)	Singvita
North-eastern hills (Purvachal) warm, per-humid, LGP >300 days (AER 17)	Khowai, Nayanpur
Eastern Coastal Plain, hot sub-humid to semi-arid, LGP 240–270 days (AER 18)	Sagar

Sehgal et al. (1992); Bhattacharyya et al. (2004).

TABLE 5.6 Description of Spatial Information in Different Agro-ecological Sub-regions (AESRs) of IGP (Level 1)

Descriptions	Soil Series
Marusthali Plains, hot hyper-arid, very low AWC, LGP <60 days (AESR 2.1)	Masitawali, Nihalkheri
Kachch Peninsula, hot hyper-arid, low AWC and LGP <60 days (AESR 2.3)	Jassi Pauwali, Nihalkheri
North Punjab Plain, Ganga-Yamuna Doab, hot semi-arid, medium AWC, LGP 90–120 days (AESR 4.1)	Fatehpur, Phaguwala, Zarifa Viran, Ghabdan
Ganga-Yamuna Doab, Rohilkhand and Avadh Plain, hot moist semi-arid, medium to high AWC, LGP 120–150 days (AESR 4.3)	Bijapur, Hirapur, Sakit
Punjab and Rohilkhand Plains, hot/dry moist sub-humid transition, medium AWC and LGP 120–150 days (AESR 9.1)	Dhoda, Jagjitpur, Bhanra, Berpura

TABLE 5.6   *(Continued)*

Descriptions	Soil Series
Rohilkhand, Avadh and south Bihar plains, hot dry sub-humid, medium to high AWC and LGP 150–180 days (AESR 9.2)	Basiaram, Itwa, Simri, Akbarpur
North Bihar and Avadh Plains, hot dry to moist sub-humid with low to medium AWC and 180–210 days LGP (AESR 13.1)	Bahraich
Foothills of Central Himalayas, warm to hot moist, high AWC and LGP 180–210 days (AESR 13.2)	Haldi
Bengal Basin and north Bihar Plains, hot moist sub-humid with medium to high AWC and LGP 210–240 days (AESR 15.1)	Sasanga, Konarpara, Hanrgram, Amarpur, Madhpur
Teesta, lower Brahmaputra Plain, hot moist humid to per-humid medium AWC and LGP 270–300 days (AESR 15.3)	Barak, Seoraguri
Foothills of Eastern Himalayas, warm to hot per-humid, low to medium AWC and LGP 270–300 days (AESR 16.1)	Singhvita
Darjeeling and Sikkim Himalayas, warm to hot per-humid, low to medium AWC and LGP 270–300 days (AESR 16.2)	Singhvita
Purvachal (Eastern Range), warm to hot per-humid, low to medium AWC and LGP >300 days (AESR 17.2)	Khowai, Nayanpur
Gangetic delta, hot moist, sub-humid to humid, medium AWC and LGP 240–270 days (AESR 18.5)	Sagar

Velayutham et al. (1999); Bhattacharyya et al. (2004).

TABLE 5.7   Description of Spatial Information in Different Agro-climatic Zones (ACZs) of IGP (Level 1)

Descriptions	Soil Series
Lower Gangetic Plains, cover 3% of TGA of the country (ACZ 3)	Sasanga, Konarpara, Hanrgram, Amarpur, Madhpur, Sagar
Middle Gangetic Plains, cover 5% of TGA of the country (ACZ 4)	Kesarganj, Bahraich, Sivapande
Upper Gangetic Plains cover 4% of TGA of the country (ACZ 5)	Basiaram, Itwa, Simri, Akbarpur, Pantnagar, Haldi
Trans-Gangetic Plains cover 4% of TGA of the country (ACZ 6)	Dhoda, Jagjitpur, Bhanra, Berpura, Nabha, Sadhu, Zarifa Viran, Fatehpur, Phaguwala, Ghabdan

TABLE 5.8 Description of Spatial Information in IGP—District Level (1:50,000 scale) (Level 2)

Particulars	Example	Remarks
Physiography	Gently sloping with slight erosion at old alluvial plain	Mainly in western part of Hooghly district of West Bengal, SISIGP, Hooghly gives complete description of physiographic classification
Slope	Gentle 1–3%	SISIGP, Hooghly district provides information on slope (quality)
Relief	20 m (above msl)	SIS, Hooghly district displays elevation of each soils of various physiography
Land Use	Mostly agriculture	Details information on vegetation is available in SISIGP, Hooghly district
Soil Order	Entisols, Inceptisols	–
Soil Series	Kota, Narahair-bati, Arandi, Satmasa, Daulatpur, Chandur, Garmandaran	–

Sarkar et al. (2001).

TABLE 5.9 Description of Spatial Information in IGP SISIGP: Watershed Level (1:2400 Scale)

Particulars	Example	Remarks[1]
Physiography	Meander flood plain	Mainly in western part of Hooghly district of West Bengal, SISIGP, Hooghly gives complete description of physiographic classification
Slope	Nearly level 0–1%	SISIGP, Chuchura Farm (Hooghly district, West Bengal) provides information on slope (quality)
Relief	5–10 m (above msl)	SISIGP, Chuchura Farm (Hooghly district, West Bengal) displays elevation of each soils of various physiography
Land Use	Mostly agriculture	Details information on vegetation is SIS, Hooghly district
Soil Order	Entisols, Inceptisols	–
Soil Series	Rabindranagar, Chuchura	–

TABLE 5.9   *(Continued)*

Particulars	Example	Remarks[1]
Soil Phases1	Rabindranagar R1A1 RmA1	Very deep, imperfectly drained, dominantly fine in control sections (25–100 cm) on A (0–1 %) slope, slightly eroded: silty clay surface
	RhA1 RjA1	Very deep, imperfectly drained, dominantly fine in control sections (25–100 cm) on A (0–1 %) slope, slightly eroded: clay surface
		Very deep, imperfectly drained, dominantly fine in control sections (25–100 cm) on A (0–1 %) slope, slightly eroded: clay loam surface
		Very deep, imperfectly drained, dominantly fine in control sections (25–100 cm) on A (0–1 %) slope, slightly eroded: Silty clay loam surface

[1]NBSS **and** LUP (1983).

TABLE 5.10   Soil Information System at Various Levels of Mapping (Level 2)

S. No.	States	Soil Map Details	Districts	No. of Reports[1]
1	Rajasthan	• Soil family (96 Nos.) • Map units (375 Nos.) • Reference (Shyampura et al., 1995)	[2]	[2]
2	Punjab	• Soil family (46 Nos.) • Map units (124 Nos) • Reference (Sidhu et al., 1995)	Patiala Hoshiapur Amritsar Jalandhar Chandigarh Ludhina	(2[a],0[b],0[c],0[d],1[e]) (1,0,0,0,1) (1,0,0,0,0) (0,1,0,0,0) (0,0,0,1,0) (0,0,0,1,0)

[1]Source: NBSS and LUP (2008).
[2] Ganganagar and Hanumangarh districts fall in IGP about which information was not available in NBSS and LUP (2008).
[a]district,[b] tahsil,[c]village,[d]farm [e] watershed reports.

TABLE 5.11   Soil Information System at Various Levels of Mapping (Level 2)

S. No.	States	Soil Map Details	Districts	No. of Reports[1]
1	Haryana	• Soil family (41 Nos.)	Karnal	(0,0,1,1,0)
		• Map units (100 Nos)	Gurgaon	(0,0,1,0,0)
		• Reference (Sachdev et al., 1995)	Hhisar	(0,0,0,1,0)
			Sirsa	(0,0,0,1,0)
2	Delhi	• Soil family (7 Nos.)	Delhi	(1,0,1,2,0)
		• Map units (31 Nos)		
		• Reference (Mahapatra et al., 1999)		

[1]Please **see** footnotes in Table 5.10.

TABLE 5.12   Soil Information System at Various Levels of Mapping (Level 2)

S. No.	States	Soil Map Details	Districts	No. of Reports[1]
1	Uttar Pradesh	• Soil family (145 Nos)	Merrut	(1,0,0,1,0)
		• Map units (321 Nos)	Sultanpur	(1,0,0,0,0)
		• Reference (Singh et al., 1999)	Etawah	(1,0,0,0,0)
			Lakhimpur	(0,1,0,0,0)
			Almora	(0,0,1,0,0)
			Barbanki	(0,0,0,2,0)
			Nanital	(0,0,0,1,0)
			Machri	(0,0,0,1,0)
			Baharaich	(0,0,0,0,1)
			Mirzapur	(0,0,0,1,0)
			Merrut cantt	(0,0,0,1,0)
			and Purazi	(0,0,0,1,0)
				(0,0,0,1,0)
				(0,0,0,1,0)
				(0,0,0,1,0)
2	Bihar	• Soil family (72 Nos.)	Champaran	(0,0,0,1,0)
		• Map units (175 Nos)	Pusa	(0,0,0,1,0)
		• Reference (Haldar et al., 1996)		(0,0,0,1,0)

[1]Please **see** footnotes in Table 5.10.

TABLE 5.13   Soil Information System at Various Levels of Mapping (Level 2)

S. No.	States	Soil Map details	Districts	No. of Reports[1]
1	West Bengal	• Soil family (56 Nos)	Bardhman	(1,0,1,0,0)
		• Map units (115 Nos)	Bankura	(1,0,1,0,0)
		• Reference (Haldar et	Hoogli	(1,0,0,1,1)
		al., 1992)	Purulia	(1,1,0,0,3)
			Birbhum	(1,0,0,0,0)
			Kuchbihar	(1,0,0,0,0)
			Jalipaipuri	(0,1,0,0,0)
			Midnapore	(0,1,1,0,0)
			Bardhman,	(0,1,0,0,0)
			Hugli and South24	(0,1,0,0,0)
			Parganas	(0,1,0,0,0)
			Haldi	(0,1,0,0,0)
			Jaldupara	(0,2,1,0,0)
			Jaldapaiguri	(0,0,1,0,0)
			Mednipur	(0,0,0,1,0)
			Asansol	(0,0,0,1,0)
			Parganas	(0,0,0,1,0)
			Kalyani	
2	Tripura	• Soil family (42 Nos)	West Tripura	(0,0,2,2,0)
		• Map units (43 Nos)	Tripura	(0,0,2,0,0)
		• Reference (Bhattacha-		(0,2,0,2,1)
		ryya et al., 1996)		

[1]Please see footnotes in Table 5.10.

5.2   APPLICATION OF SOIL INFORMATION SYSTEM

Soil information system has been used to estimate soil organic carbon stocks as shown in Table 5.14. Soil information system has also been found to be useful to assess soil quality (Bhattacharyya and Mandal, 2009) (Fig. 5.8). The other utilities of SIS has been detailed elsewhere (Bhattacharyya et al., 2009). Soil inventory encompasses database for crop history as was used during execution of Global Environmental Facility sponsored Soil Organic Carbon project. The database was developed for crops grown extensively in various parts of the IGP in terms of tillage, fertilization, irrigation and harvest various crops including agriculture and horticulture (Bhattacharyya et al., 2007). The soil information system and the crop history data were used for GEFSOC modeling system for building land management tree, diagrams building different data layers and model run file, analysis of data to arrive at the final output in the form of organic C stock estimates. An example of management tree is shown in Fig. 5.9. A comparison of estimated vs. modeled SOC stock is shown in Fig. 5.10. The data indicate that IGP might have reached a dynamic equilibrium after 30–40 years of Green Revolution (Bhattacharyya et al., 2007).

TABLE 5.14   SOC Stock

Soil/System	Method	Year of Estimate	SOC Stock (Pg)
Soil Series of IGP	Soil Survey	1980	0.66
Soil Series of IGP	Soil Survey (Resampling)	2005	0.88
Soil Series of IGP	Secondary data extrapolation	1990	0.572–0.587
IPCC	Soil, climate and land use classification	2000	0.97
Century C model	Simulation	2000	1.44

Source: Bhattacharyya et al. (2007).

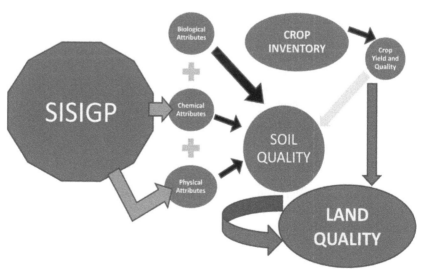

FIGURE 5.8   Soil Information System vis-à-vis Soil and Land Quality.

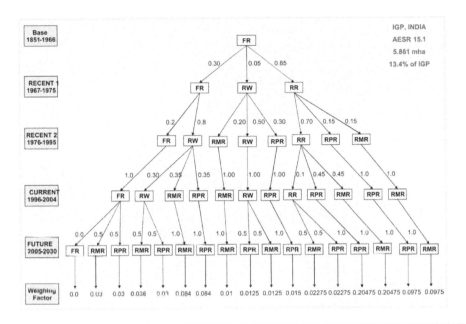

Management sequence chains for IGP, India (AESR 15.1). System Key: FR=Fallow-Rice; RW=Rice-Wheat; RR=Rice-Rice; RMR=Rice-Mustard-Rice; RPR=Rice-Potato-Rice; Number next to each trajectory path indicate the rank, which indicates the order in which acreages associated with specific systems transition into sequences that follow in the chain. The number 1 is the highest rank.

FIGURE 5.9 Management Tree Diagram for AESR 15.1 used for GEFSOC model.

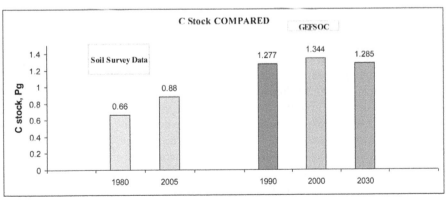

FIGURE 5.10 Comparison of Estimated vs Modeled SOC Stocks.

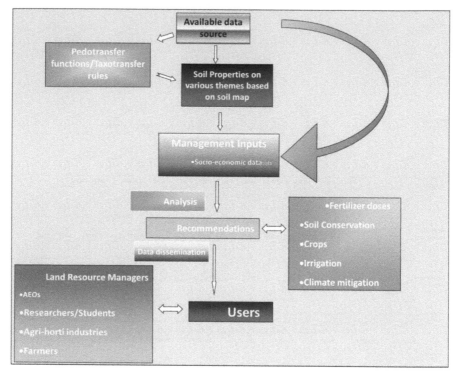

FIGURE 5.11   SISIGP (digital) Concepts for Data Framing for end Users.

5.3   CONCLUSION

Soil information system can thus be an extremely helpful tool for data framing for end users including research managers, extension people, researchers and students for recommending fertilizer doses, soil conservation, crop management and planning land use options, irrigation scheduling and climate mitigation (Fig. 5.11). In view of the global changing scenario the need of the hour is to produce a fresh group of earth scientists with specialization in soil and crop science, geology and geography with appreciable knowledge in GIS and information technology software. They should be equipped to deal with data storage, and retrieval in a user-friendly mode for management recommendations, so that issues like land degradation, biodiversity, food security and climate change can be addressed adequately.

KEYWORDS

- **GEFSOC modeling system**
- **Geo-Referencing Soil Information System**
- **Geographic Information System**
- **Global Environmental Facility**
- **Indo-Gangetic Plains**
- **Soil and Terrain Digital Database**
- **Soil Information System**

REFERENCES

Abrol, I. P., Gupta, R. K., (1998). Indo-Gangetic Plains – Issues of changing land use. LUCC Newsl., March 1998, No. 3.

Batjes, N. H., Al-Adamat, R., Bhattacharyya, T., Bernoux, M., Cerri, C. E. P., Gicheru, P., Kamoni, P., Milne, E., Pal, D. K., Rawajfih, Z. (2007). Preparation of consistent soil datasets for modeling purposes: Secondary SOTER data for four case study areas. Agriculture, Ecosystems and Environment 122, 26 34.

Bhandari, A. L., Ladha, J. K., Pathak, H., Padre, A. T., Dawe, D., Gupta, R. K. (2002). Yield and soil nutrient changes in a long-term rice-wheat rotation in India. Soil Sci. Soc. Am. J. 66, 162–170.

Bhattacharjee, J. C., Roychaudhury, C., Landey, R. J., Pande, S. (1982). Bioclimatic Analysis of India. NBSSLUP Bulletin 17, NBSS&LUP, Nagpur 21pp+map.

Bhattacharyya, T., Easter, M., Paustian, K., Killian, K., Williams, S., Ray, S. K., Chandran, P., Durge, S. L., Pal, D. K., Gajbhiye, K. S., Milne, E., Singh B. (2005). Description of newly added crop database and modified crop files: Indo-Gangetic plains, India case study. Special publication for: Assessment of Soil Organic Carbon Stocks and Change at National Scale. NBSS & LUP, Nagpur, India, 196 pp.

Bhattacharyya, T., Mandal, B. (2009). Soil Information System of the Indo-Gangetic Plains for resource management. Paper presented in ISSS special session on Land Use Planning during Platinum Jubilee Celebration at IARI New Delhi, December 22–25.

Bhattacharyya, T., Pal, D. K., Chandran, P., Mandal, C., Ray, S. K., Gupta, R. K., Gajbhiye, K. S. (2004). Managing Soil Carbon Stocks in the Indo-Gangetic Plains, India. Rice-Wheat Consortium for the Indo- Gangetic Plains, New Delhi – 110 012, India. Pp. 44.

Bhattacharyya, T., Pal, D. K., Easter, M., Batjes, N. H., Milne, E., Gajbhiye, K. S., Chandran, P., Ray, S. K., Mandal, C., Paustian, K., Williams, S., Killian, K., Coleman, K., Falloon, P., Powlson, D. S. (2007). Modelled soil organic carbon stocks and changes in the Indo-Gangetic Plains, India from 1980 to 2030. Agriculture Ecosystems and Environment 122 (1), 84–94.

Bhattacharyya, T., Sarkar, Dipak, Pal, D. K., Mandal, C., Baruah, U., Telpande, B., Vaidaya, P. H. (2009). Soil Information System for resource management- Tripura as a case study, Current Science 99, 1208–1217.

Bhattacharyya, T., Sehgal, J., Sarkar, D. (1996). Soils of Tripura for optimizing land use: their kinds, distribution and suitability for major field crops and rubber. NBSS Publ. 65 b (Soils of India series 6). NBSS & LUP, Nagpur, India 14 p.

Chandran, P., Mandal, C., Bhattacharyya, T., Dipak Sarkar, Pal, D. K., Ray, S. K., Jagdish Prasad, Tiwary P., Srivastava, R., Lokhande, M., Wadhai, Dongre, V., Dijkshoorn J. A., Batjes N. H., Brindaban, P. S. (2012). Soil and Terrain Information System for Resource Planning of the Indo-Gangetic Plains, India. Paper presented at National Seminar on GEOS-2012 held during January 18–20, 2012 at Karnataka State Remote Sensing Application Centre, Bangalore, January-2012, 25–35.

Govinda Rajan, S. V. (1971). Soil map of India. In Review of Soil Reserach in India (J. S. Kanwar and S. P. Raychaudhuri, Eds), Indian Society of Soil Science, New Delhi.

Gupta, R. K. (2003). The rice-wheat consortium for the Indo-Gangetic Plains: vision and management structure. In: RWC-CIMMYT. (Ed.), Addressing Resource Conservation Issues in Rice-Wheat Systems of South Asia: A Resource Book. RWC-CIMMYT, New Delhi, India, ISBN 81- 88572–00–4, pp. 1–7.

Haldar, A. K., Srivastava, R., Thampi, C. J., Sarkar, D., Singh, D. S., Sehgal, J., Velayutham, M. (1996). Soils of Bihar for Optimizing Land Use. NBSSLUP Publication 50b (Soils of India Series), National Bureau of Soil Survey and Land Use Planning, Nagpur, India. 70 pp. + 4-sheet soil map (1:500,000 scale).

Haldar, A. K., Thampi, C. J., Sehgal, J. (1992). Soils of West Bengal for Optimizing Land Use. NBSSLUP Publication 27b. (Soils of India Series). National Bureau of Soil Survey & Land Use Planning, Nagpur (India) 48p. + 4 Sheet soil map.

Mahapatra, S. K., Rana, K. P. C., Tersem Lal, Velayutham, M. (1999). Soils of Delhi for Optimizing Land Use. NBSS&LUP Publication 72b (Soils of India series). National Bureau of Soil Survey and Land Use Planning, Nagpur, India 73pp + 4 sheets of Soil Map on 1: 125, 000 scale.

Mandal, C., Mandal, D. K., Bhattacharyya, T., Sarkar, D., Pal, D. K., Prasad, J., et al. (2013). Revised agro-ecological Sub Region for Land Use Planning, Working Report No.12, NAIP Component – 4 Project on "Georeferenced Soil Information System for Land Use Planning and Monitoring Soil and Land Quality for Agriculture." NBSS&LUP, Nagpur. p.50.

NBSS&LUP (1983). Soil Survey Report of Chuchura Farm, district Hooghly, West Bengal, NBSS&LUP, Regional center Calcutta, Report No 448. p. 12.

NBSS&LUP (1985). Soils of India (Suborder association 1:7 M), NBSS&LUP, Nagpur.

NBSS&LUP (2008). Soil Survey Reports of Research Farms/Villages/Tehsils/Districts/Watersheds/ Catchments/Command Areas/Atlases (from 1959 to 2005) NBSS&LUP, Nagpur.

NBSS&LUP. (2002). Soils of India (Family association 1:1 M), NBSS&LUP, Nagpur.

NRCS (1996). Global Soil Regions. United States Department of Agriculture, Natural Resources Conservation Service, Soil Survey Division, World Soil Resources.

Ray, S. K., Chandran, P., Bhattacharyya, T., Durge, S. L., Mandal, C., Sarkar, D., Sahoo, A. K., Singh, S. P., Jagat Ram, Ram Gopal, Pal, D. K., Gajbhiye, K. S., Milne, E., Singh, B., Aurangabadkar, B. (2005). Benchmark Soil Series of the Indo-Gangetic Plains (IGP), India. Special publication for Assessment of Soil Organic Carbon Stocks and Change at National Scale. NBSS&LUP, India. p.186.

Sachdev, C. B., Tarsem Lal, Rana, K. P. C., Sehgal, J. (1995). Soils of Haryana for Optimizing Land Use Planning, Nagpur, India, 59 pp +2 sheet soil map (1:500,000 scale).

Sarkar, D., Das, T. H., Chattopadhyay, T., Velayutham, M. (2001). Soils of Hooghly district for optimizing land use, NBSS&LUP Publication 88, NBSS&LUp, Nagpur, 91 p.

Sehgal, J., Mandal, D. K., Mandal, C., Vadivelu, S. (1992). Agro-Ecological Regions of India, Second Edition. Technical Bulletin No. 24, NBSS&LUP, Nagpur, India, 130p.

Shyampura, R. L., Sehgal, J. (1995). Soils of Rajasthan for Optimizing Land Use planning. Nagpur, India, +6 sheet soil map (1:500,000 scale).

Sidhu, G. S., Walia, C. S., Tarsem Lal, Rana, K. P. C., Sehgal, J. (1995). Soils of Punjab for Optimizing Land Use, NBSS&LUP Publication 45 (Soils of India Series 4). National Bureau of Soil Survey and Land Use Planning, Nagpur, India, 75p +2 sheet soil map (1:500,000 scale).

Singh, S. P., Rana, K. P. C., Sehgal, J., Velayutham, M. (1999). Soils of Uttar Pradesh for Optimizing Land Use. NBSS Publ. (Soils of India series). National Bureau of Soil Survey and Land Use Planning, Nagpur, India (in press).

Velayutham, M., Mandal, D. K., Mandal, C., Sehgal, J. (1999). Agro-ecological Subregion of India for Planning and Development. NBSS Publication 35, 372p. NBSS&LUP, Nagpur, India.

CHAPTER 6

RECENT ADVANCES IN THE ASSESSMENT OF DEGRADED LANDS FOR THEIR MANAGEMENT

RAJEEV SRIVASTAVA, DIPAK SARKAR, and G. P. OBI REDDY[1]

CONTENTS

[1]National Bureau of Soil Survey and Land Use Planning, Amravati Road, Nagpur–440 033

ABSTRACT

Shrinking land resources coupled with accelerated rate of land degradation and ever increasing population driven food demand has drawn attention of the administrators, planners, scientific communities and other stakeholders towards the need of restoring the existing degraded soils so as to use the same for increasing productivity. Human induced land degradation due to unsustainable land use and inappropriate land management practices is of major concern. Recent estimates documented 120.72 M ha of the country as degraded/wastelands of which water erosion being the major one (73.27 mha). The chapter discusses the applicability of recent advances in satellite remote sensing, diffused reflectance spectroscopy (DRS) and GIS technologies in developing a comprehensive geo-referenced land information system as well as strong database for real time monitoring of soil quality. This may serve as an important linkage for environmental monitoring, accounting, and impact assessment to land degradation.

6.1   INTRODUCTION

Land degradation is an important global issue because of its adverse impact agronomic productivity, the environment, and its effect on food security and the quality of life (Eswaran et al., 2004). In India, about 70% of the population is entirely dependent on agriculture as their livelihood support system. The vast majority of Indian farmers are small and marginal. Their farm size is decreasing further due to population growth.

With the advent of green revolution technology, no doubt food productivity has increased many times. However, non/partial adoption of package of practices, high population pressure, erratic behavior of monsoon and miseries imposed by nature induced various kind of degradation (Singh and Singh, 2010). Intensive farming practices without adequate nutrient replenishment, have virtually mined nutrients from the soil. Due to heavy use of fertilizers, excess nitrates have leached into groundwater and contamination of groundwater with nitrates has increased dramatically (Datta et al., 1995). It is becoming increasingly recognized that agriculture, particularly the increased use of marginal lands, is an important cause of environmental degradation. Increasingly less remunerative subsistence farming systems lead the cultivators to leave the land uncultivated, which then tends to become barren, and ultimately result in shifting the land resources from agriculture to other uses including industry, which bring environmental degradation. All of these factors combined with increased rate of land degradation are contributing towards decline in agricultural productivity leading to food insecurity.

Since land resources are finite, requisite measures are required to manage the natural resources so that they are neither degraded nor depleted and ensure a sustained production for future generations. For sustainable development, available land resources need to be used based on their potential and limitations. It implies that the productivity of existing degraded soils needs to be restored and that of normal cultivated lands are to be improved.

6.2 CAUSES OF LAND DEGRADATION

Degradation of land, in general, is a consequence of either natural hazards or anthropogenic causes. Natural hazards are the environmental conditions, which lead to high susceptibility to erosion such as high intensity storms on steep slopes and soils having less resistance to water erosion, high speed winds, soil fertility decline due to strong leaching in humid climates, acidity or loss of nutrients, waterlogging, etc. The anthropogenic causes are human induced, which result from unsustainable land use and inappropriate land management practices such as deforestation and over-exploitation of vegetation, overgrazing, cultivation on steep slopes and marginal/ fragile lands without adoption of soil conservation measures, shifting cultivation, improper crop rotations, imbalanced fertilizer use or excessive use of agro-chemicals, overexploitation of groundwater and improper management of canal water.

In the context of productivity, land degradation results from a mismatch between land quality and land use (Beinroth et al., 1994). Mechanisms that initiate land degradation include physical, chemical, and biological processes (Lal, 1994). Important among physical processes are a decline in soil structure leading to crusting, compaction, erosion, desertification, anaerobism, environmental pollution, and unsustainable use of natural resources. Significant chemical processes include acidification, leaching, salinization, decrease in cation retention capacity, and fertility depletion. Biological processes include reduction in total and biomass carbon, and decline in land biodiversity. The latter comprises important concerns related to eutrophication of surface water, contamination of groundwater, and emissions of trace gases (CO_2, CH_4, N_2O, NO_x) from terrestrial/aquatic ecosystems to the atmosphere (Eswaran et al., 2001).

6.3 LAND DEGRADATION STATISTICS

Because of different definitions and terminology, and different methods adopted by different agencies, there also exists a large variation in the available statistics on the extent and rate of land degradation. Further, most statistics refer to the risks of degradation or desertification (based on climatic factors and land use) rather than to the current state of the land (Eswaran et al., 2001).

It is estimated that about 2 billion ha area in the world that was once biologically productive is now affected by various forms of land degradation (Oldeman, 1991). About 5–7 million ha of arable land of the world is lost annually through land degradation (Lal and Stewart, 1990). Globally, land degradation affects about one-sixth of the world's population, 70% of all dry lands (about 3.6 billion ha) and one-quarter of the total land area of the world. The continental percentage of land degradation is highest in Asia (37%) followed by Africa (25%), South America (14%), Europe (11%), North America (4%) and Central America (3%), the world total being 15%.

In India, the estimates of land degradation by different agencies vary widely from about 53 M ha to 188 M ha, mainly due to different approaches adopted in defining degraded lands and/or differentiating criteria used (Sharda, 2011). The earliest assessment of the area affected by the land degradation was made by the National Commission on Agriculture (NCA, 1976) at 148 M ha, followed by 175 M ha by the Ministry of Agriculture (Soil and Water Conservation Division). National Remote Sensing Agency (NRSA, 1985) mapped the wastelands using satellite data of 1980–82 and reported that 53.3 M ha, representing 16.2% of the total geographical area is under different categories of wastelands. The National Wasteland Development Board (1985) estimated an area of 123 M ha under wastelands. The NBSS&LUP estimates projected an area of 187.7 M ha (Sehgal and Abrol, 1994) as degraded lands in 1994 following Global Assessment of Soil Degradation (GLASOD) methodology (Oldeman, 1988), and revised it to 147 M ha in 2004. Recent estimates of NBSS&LUP generated based on harmonization of degradation/wasteland datasets of different agencies indicate that 120.72 M ha (Table 6.1) as degraded/wastelands (ICAR-NAAS, 2010).

TABLE 6.1   Harmonized Area Statistics of Degraded and Wastelands of India

S. No	Type of Degradation	Arable land (in Mha)	Open forest (<40% Canopy) (in Mha)
1	Water erosion (>10 t/ha/yr)	73.27	9.30
2	Wind erosion (Aeolian)**	12.40	-
	Sub total	85.67	9.30
3	Chemical degradation		
	a) Exclusively salt affected soils	5.44	-
	b) Salt-affected and water eroded soils	1.20	0.10
	c) Exclusively acidic soils (pH< 5.5)	5.09	-
	d) Acidic (pH < 5.5) and water eroded soils	5.72	7.13
	Sub total	17.45	7.23
4	Physical degradation		
	a) Mining and industrial waste	0.19	
	b) Water logging (permanent) (water table within 2 mts depth)*	0.88	
	Sub total	1.07	
	Total	104.19	16.53
	Grand total (Arable land and Open forest)		120.72

Source: ICAR-NAAS (2010).

A major shortcoming of the available statistics on land degradation is the lack of cause–effect relationship between severity of degradation and productivity. Criteria for designating different classes of land degradation (e.g., low, moderate, high) are generally based on land properties rather than their impact on productivity. In fact, assessing the productivity effects of land degradation is a challenging task (Lal, 1998). Assessing the productivity effects of land degradation requires a thorough understanding of the processes involved at the soil–plant–atmosphere continuum. These processes are influenced strongly by land use and management.

6.4   TOOLS OF LAND DEGRADATION ASSESSMENT

Land degradation is also a complex issue and degradation processes differ greatly. A suitable method for assessment in one situation may not be suitable in another situation. In most of the soil degradation studies conducted globally or at national level, expert assessment/opinion has been the main method used. GLASOD is subjective and is based on expert opinion only. GLASOD methodology distinguishes four human-induced processes of soil degradation: water and wind erosion, plus chemical and physical degradation (Oldeman et al., 1990).

Remote sensing, field monitoring, observation of productivity changes and land users' opinion and modeling are other methods that are also being used in land degradation assessments (FAO, 2001, Van Lynden and Kuhlmann, 2002). Remote sensing has the greatest comparative advantage when the scale is small, because it can provide data for a large area at one time (Van Lynden and Kuhlmann, 2002), and it is, in principle, an ideal methodology for regional or global degradation assessments. The main problem with the method is that the data should not be used as such, but should be accompanied by adequate ground data in order to obtain reliable estimates.

Recent research (Brown et al., 2005; Shepherd and Walsh, 2002; Summers et al., 2011; Viscarra Rossel et al., 2006) has demonstrated the ability of DRS to provide nondestructive rapid prediction of various soil quality attributes (i.e., physical, chemical and biological properties) in the laboratory. DRS of soil is an integrative property, responding to mineral composition, iron oxides, organic matter, water (soil functional hydration, hygroscopic and free), carbonates, soluble salts and particle size distribution. These properties largely determine functional capacity (for example, ability to support plant growth, hydraulic regulation, resistance to erosion, engineering properties). DRS is based on the use of calibrations, coupled with chemometrics techniques, which use absorbances at many wavelengths to predict particular properties of a sample (Shepherd and Walsh, 2002). For this reason, DRS could be used as a powerful tool for assessment and monitoring of soil quality/soil degradation.

The study carried out at NBSS&LUP on the prediction of soil properties of shrink-swell soils of Central India revealed that soil variables viz. soil pH, soil organic carbon (SOC), cation exchange capacity and clay could be predicted from diffused visible near-infrared (VNIR) soil reflectance data (Srivastava et al., 2004). Under the NAIP project "Development of Spectral Reflectance Methods and Low Cost Sensors For Real-Time Application of Variable Rate Inputs in Precision Farming" reflectance spectral models have been developed for prediction of soil properties (NBSS staff 2012, NAIP 2012) of the IGP from soil reflectance data (350–2500 nm) using partial least-square (PLS) regression technique and validated. The developed model showed very good prediction ($r^2 > 0.8$), of soil organic carbon (OC), electrical conductivity (ECe), saturation extract Na^+, saturation extract Ca^{2+} and Mg^{2+}, saturation extract cl^- and saturation extract SO_4^{2-} in the validation datasets.

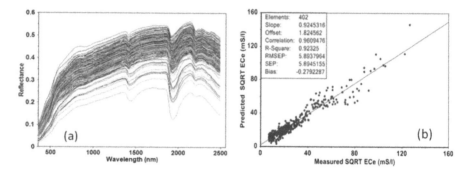

FIGURE 6.1   Laboratory Measured Soil Reflectance Spectra of Salt-affected Area of Haryana (a) and Scatter-plot of Measured and Predicted Values of ECe (mS/m) from Soil Reflectance Data (b).

The joint study conducted by the NBSS & LUP, Nagpur and SAC, Ahmedabad on applicability of Hyperion data (Hyperspectral) on soil fertility zonation in Buldhana district, Maharashtra indicated good correlation between hyperion reflectance data and SOC content, thus can be used for spatial mapping of SOC (Maji et al., 2010).

Shepherd and Walsh (2007) proposed a diagnostic surveillance framework modeled on medical diagnostic approach for evidence-based management of agriculture and environment. They observed that infrared spectroscopy can play a pivotal role in making the surveillance framework operational by providing a rapid, low cost and highly reproducible diagnostic screening tool. Spectral indicators would help to quantify risk factors associated with problems and assess intervention impacts.

6.5 CONCLUSION

Land degradation results from mismanagement of land and thus it affects both the natural ecosystem and the human social system. Food security, environmental balance, and land degradation are strongly interlinked and each must be addressed in the context of the other to have measurable impact. Land degradation is as much a socioeconomic problem as it is a biophysical problem. Land use must match land quality; appropriate national policies should be implemented to ensure this occurs to reduce land degradation. The focus of agricultural research should shift from increasing productivity to enhancing sustainability, recognizing that land degradation caused by agriculture can be minimized and made compatible with the environment. A comprehensive geo-referenced land information system needs to be developed that link environmental monitoring, accounting, and impact assessment to land degradation. The DRS may help in developing a strong database of soil quality, which can be used for the assessment and monitoring of the degraded areas.

Degradation is a slow imperceptible process and many people are not aware that their land is degrading. Creating greater awareness of the perils of land degradation in society and political leadership are important steps in the challenge of reducing degradation.

KEYWORDS

- **Chemometrics Techniques**
- **Diffused Reflectance Spectroscopy**
- **GLASOD**
- **Land Degradation Assessment Tools**
- **Soil–Plant–Atmosphere Continuum**
- **Water/Wind Erosion**

REFERENCES

Beinroth, F. H., Eswaran, H. Reich, P. F., Van Den Berg, E. (1994). Land stresses in agroecosystems. In: stressed Ecosystem and sustainable Agriculture, eds. S. M. Virmani, J. C. Katyal, H. Eswaran, I. P. Abrol, New Delhi: Oxford and IBH.

Brown, D. J., Bricklemyer, R. S. Miller, P. R. (2005). Validation requirements for diffuse reflectance soil characterization models with a case study of VNIR soil C prediction in Montana. Geoderma, 129, pp. 251–267.

Datta, P. S., Deb, D. L. Tyagi. S. K. (1997). Assessment of groundwater contamination from fertilizers in the Delhi area based on 180, NO_3^- and K^+ composition. Journal of Contaminant Hydrology, 27, 249–262.

Eswaran, H., Lal, R., Reich, P. F. (2001). Land degradation: an overview. In: Bridges, E. M., Hannam, I. D., Oldeman, L. R., Pening de Vries, F. W. T., Scherr, S. J., Sompatpanit, S. (eds.). Responses to Land Degradation. Proc. 2nd. International Conference on Land Degradation and Desertification, Khon Kaen, Thailand. Oxford Press, New Delhi, India.

FAO, (2001). Report on the International Workshop on the Land Degradation Assessment in Drylands Initiative (LADA), 5–8 December 2001, FAO, Rome.

ICAR-NAAS (2010). Degraded and Wastelands of India: Status and Spatial Distribution, Indian Council of Agricultural Research-National Academay Agricultural Sciences, New Delhi, P 158.

Islam, K., Singh, B. McBratney, A., (2003). Simultaneous estimation of several soil properties by ultra-violet, visible, and near-infrared reflectance spectroscopy. Australian Journal of Soil Research, 41, 1101–1114.

Lal, R. (1994). Tillage effects on soil degradation, soil resilience, soil quality and sustainability. Soil Tillage research, 27, 1–8.

Lal, R. (1998). Soil erosion impact on agronomic productivity and environment quality. Critical Reviews in Plant Sciences, 17, 319–464.

Lal, R., Stewart, B. A. (1990). Soil degradation: A global threat. Advances in Soil Scence., 11: XIII–XVII.

Maji, A. K., Srivastava, Rajeev, Sarkar, Dipak Mehta, R. L., Ray, S. S. (2010). Soils variability mapping and fertility zonation using Hyperspectral data. National Bureau of Soil Survey and Land Use Planning (ICAR), Nagpur and Space Application center (DOS), Ahmedabad. 39p.

NAIP. (2012). Annual Report 2011–12: National Agricultural Innovation Project ICAR, New Delhi.

NBSS Staff (2012). Annual Report, 2011–12, NBSS&LUP Publ., Nagpur-440 033, India (unpublished).

NBSS&LUP (2005). Reflectance libraries for development of soil sensor for periodic assessment of state of soil resources. – NATP Project Report (NBSS No. 835).

NCA, (1976). Report of the National Commission on Agriculture: Part V, IX and Abridged. Ministry of Agriculture and Irrigation, New Delhi.

NRSA, (2005). Wasteland Atlas of India. Ministry of Rural Development and NRSA Publ., NRSA, Hyderabad.

Oldeman, L. R. (1991). Global extent of soil degradation. Bi-annual Report, ISRIC, Wageningen. The Netherlands, pp. 19–35.

Oldeman, L. R. (ed.) (1988). Guidelines for general assessment of the Status of Human-induced Soil Degradation. Working paper and Preprint no. 88 (4), ISRIC, Wageningen.

Sehgal, J., Abrol, I. P. (1994). Soil degradation in India: status and impact. New Delhi: Oxford and IBH. 80 pp.

Sharda, V. N. (2011). Strategies for arresting land degradation in India. In: Dipak Sarkar, A. K. Azad, Singh, S. K., Akter, N. (eds.), Strategies for arresting land degradation in South Asian countries, SAARC Centre, Dhaka.

Shepherd, K. D., Walsh, M. G. (2002). Development of reflectance spectral libraries for characterization of soil properties. Soil Science Society of America Journal, 66, pp. 988–998.

Singh, A. K., Singh, S. K. (2010). Keynote paper on strategies for arresting land degradation in South Asian countries. In: Dipak Sarkar, A. K. Azad, S. K. Singh and N. Akter (eds.), Strategies for arresting land degradation in South Asian countries, SAARC Centre, Dhaka.

Srivastava, R., Jagdish Prasad, Saxena, R. K., (2004). Spectral reflectance properties of some shrink-swell soils of Central India as influenced by soil properties. Agropedology, 14, pp. 45- 54.

Summers, D. Lewis, M. Ostendorf, B. Chittleborough, D., (2011). Visible near-infrared reflectance spectroscopy as a predictive indicator of soil properties. Ecological Indicators, 11, pp. 123–131.

Van Lynden, G. W. J. Kuhlmann, T. (2002). Review of Degradation Assessment Methods. LADA (Land Degradation Assessment in Dryland Areas). Draft, September 2002. 53p.

Viscarra Rossel, R. A., Walvoort, D. J. J., McBratney, A. B., Janik, L. J. Skjemstad, J. O. (2006). Visible, near infrared, mid infrared or combined diffuse reflectance spectroscopy for simultaneous assessment of various soil properties. Geoderma, 131, 59–75.

PART II

APPLICATION OF GEO-INFORMATICS IN LAND USE
PLANNING

CHAPTER 7

GEOSPATIAL TECHNOLOGIES FOR LAND USE PLANNING, SUSTAINABLE LAND RESOURCE MANAGEMENT AND FOOD SECURITY

G. P. OBI REDDY[1] and DIPAK SARKAR[1]

CONTENTS

National Bureau of Soil Survey & Land Use Planning, Amravati Road, Nagpur – 440 033, India

ABSTRACT

Information on the nature, extent, and spatial distribution of land resources is prerequisite for ensuring optimal use of land resources for sustainable development, poverty alleviation and food security. Geospatial technologies like photogrammetry, satellite remote sensing, Global Positioning System (GPS) and Geographic Information System (GIS) play a vital role in providing precise information on nature, extent, and spatial distribution of land resources to assess their potentials and limitations for planning, monitoring and management towards sustainable development. The applications of space technology are rapidly advancing in land resources inventory, mapping and generation of databases on a regular basis for better planning, management, monitoring and implementing the land use plans at large scales. GIS techniques are playing an increasing role in facilitating integration of multilayer spatial information with statistical attribute data to arrive at alternate developmental scenarios in sustainable management of land resources. In this chapter, geospatial technologies in land resource inventory, mapping, evaluation and management including digital terrain database generation, hydrological analysis, landform mapping, soil resource inventory, land use systems analysis, soil-site suitability evaluation, soil loss assessment, watershed prioritization, carrying capacity assessment and development of spatial decision support systems for sustainable land resource management for land use planning and food security have been discussed.

7.1 INTRODUCTION

Ever increasing population with increasing life expectancy is putting tremendous pressure on the finite land resources to provide adequate food grains, water, energy, living space and other necessities of the life in developing countries like India. In order to improve the efficiency and sustainability of land resources, knowledge and information on the land resources and their distribution in space and time are essential (Young, 1998). Information on the nature, extent, and spatial distribution of land resources is prerequisite for ensuring optimal use of land resources for sustainable development, poverty alleviation and food security. Soil, landform, vegetation and climate, or combinations of these ecosystem components, have been used as indicators of a variety of biophysical characteristics in the different land evaluation, or assessment, systems developed and applied around the world (Westman, 1985). These biophysical characteristics can be used to predict the suitability of land for different uses, or as indicators of the vulnerability of the land. In land resources appraisal, evaluation, planning and management at microlevel, watershed is considered as an appropriate physical unit. Watershed development program is one of the major initiatives in the country towards conservation of soil and water resources in the rainfed area for enhancing agricultural production, and to ensure livelihood security to rural people. Geospatial technologies have immense potential in inventory, characterization and sustainable management of land resource at watershed level.

Geospatial technologies (photogrammetry, satellite remote sensing, Global Positioning System and Geographic Information System) play a vital role in providing precise information on nature, extent, and spatial distribution of land resources to assess their potentials and limitations for planning, monitoring and management towards sustainable development. High resolution remotely sensed data provide an unparalleled view of the Earth for studies that require synoptic or periodic observations such as inventory, surveying, mapping and monitoring in land resources, land use/land cover and environment. The applications of space technology are rapidly advancing in land resources inventory, mapping and generation of databases on a regular basis for better planning, management, monitoring and implementing the land use plans at large scales. Satellite remote sensing data from LANDSAT-ETM+, IRS-IC, IRS-ID, IRS-P6, Cartosat-1, Cartosat-II, Quickbird and Google are available for generation of spatial database on land resources for various applications. GIS techniques are playing an increasing role in facilitating integration of multi-layer spatial information with statistical attribute data to arrive at alternate developmental scenarios in sustainable management of land resources. The integration of spatial data and their combined analysis could be performed through GIS and simple database query systems to complex analysis and decision support systems could be developed for effective land resource management at watershed level. In this chapter, the scope and potential applications of geospatial technologies in inventory, mapping and management of land resource for land use planning and food security have been discussed with few examples at watershed level.

7.2   GEOSPATIAL TECHNOLOGIES IN LAND RESOURCE INVENTORY AND MAPPING

Several geospatial technologies including remote sensing from terrestrial, aerial, and satellite platforms with various sensors. GPS and GIS are actively being used in land resource inventory and mapping. Geospatial technologies in land resource inventory and mapping particularly in digital terrain database generation, hydrological analysis, landform mapping, soil resource inventory and land use systems analysis have been discussed below.

7.2.1   DIGITAL TERRAIN DATABASE GENERATION

Geospatial technology play significant role in development of digital terrain database and digital terrain modeling at watershed level in replacing the qualitative and nominal characterization of topography. The availability of satellite based new topographic datasets has opened new venues for hydrologic and geomorphologic studies including analysis of surface morphology (Frankel and Dolan, 2007). The digital terrain database shown its comparative advantages in quantitative measurement of elevation, enables to derive any other terrain attribute quantitatively, enables to visualize topography in more realistic way than ever before, and enables to analyze various geomorphometric parameters digitally (Li et al., 2005; Moore et al., 1993; Wilson and Gallant, 2000a). Digital elevation models (DEM) provide good terrain representation from which the watersheds can be derived automatically using GIS technology. Nowadays, several radar satellite based DEMs like SRTM 90 m, ASTER 30 m and Cartosat DEM 30 m are available for the public but the choice of the DEM, which better suits the target of the study is crucial. It further provides the possibility of deriving indices that can be used as indicators for environmental processes (Pike, 1988; Wilson and Gallant, 2000b). The other paradigms such as pedometrics and digital soil mapping are widening the scope and applicability of DEMs (McBratney et al., 2003). Seamless mosaic of Shuttle Radar Topographic Mission (SRTM) digital elevation data (90 m) for India has been developed at NBSS&LUP to analyze and characterize the selected geomorphological parameters namely slope, aspect, hill shade, plane curvature, profile curvature, total curvature, flow direction, flow accumulation and topographic wetness index at state and agroecological regions of India (Reddy et al., 2011). Digital elevation model and slope maps of Maharashtra derived from SRTM data are shown in Fig. 7.1.

FIGURE 7.1   Digital Elevation Model and Slope Maps of Maharashtra Derived from SRTM data.

7.2.2   HYDROLOGICAL ANALYSIS

Accurate delineation of drainage networks is a prerequisite for many natural resource management issues (Liu and Zhang, 2010; Paik, 2008). Due to the increasing availability of grid DEMs, numerous research studies have been carried out to automate the extraction of drainage networks (Bertolo, 2000; Moore et al., 1991; Tribe, 1992). The remote sensing and GIS tools have opened new paths in large-scale hydrological studies. The high resolution DEMs have immense potential in generation of detailed drainage pattern, 3-D perspective views, viewshed analysis and other terrain variables, which can be used as a input in land resource inventory, mapping and management at watershed level. Temporal data from remote sensing enables identification of groundwater aquifers and assessment of their changes along with the land use changes, whereas, GIS enables integration of multithematic data. Analysis of surface and subsurface hydrological parameters through analysis of remotely sensed data in conjunction with field surveys provides a scope to identify hydro-geomorphological units. It can be carried out in GIS through database generation on drainage, landforms, lithology, land use/land cover and lineament thematic parameters from data and their overlay analysis (Krishnamurthy and Srinivas, 1995; Reddy et al., 1999). The delineated hydro-geomorphological units at watershed level can be analyzed to assess their groundwater potential/deficit conditions. Reddy et al., (2006) performed multicriteria overlay analysis in GIS considering the IRS-ID LISS-III data, slope, geological and landform units, drainage density, soil depth, lineament density and land use/land cover layers to delineate distinct groundwater potential zones in Tirora tehsil, Maharashtra. However, the necessary field verification in conjunction with available collateral data is essential for site specific groundwater exploration in the tehsil. IRS-ID LISS-III data and groundwater potential zones of Tirora tehsil, Gondia district, Maharahstra are shown in Fig. 7.2.

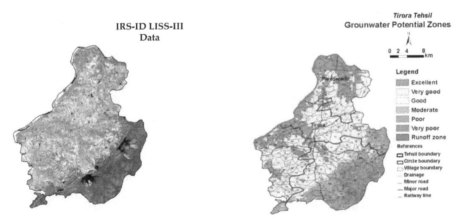

FIGURE 7.2   IRS-ID LISS-III Data and Groundwater Potential Zones of Tirora Tehsil, Gondia District, Maharashtra.

7.2.3   TERRAIN ANALYSIS AND LANDFORM MAPPING

Digital terrain modeling has many applications in hydrological and geomorphological studies (Moore et al., 1991). Delineation of terrain parameters, such as drainage network and watershed boundaries from collateral data and remotely sensed data crucially depends on generation of an accurate DEM. DEM have been used in terrain analysis and it involves both interactive and interpretative methods, requiring repeated visualization of the resulting classified maps and adjustment of classification parameters. Moreover, it is specific to physiographic units having similar topographic and geologic characteristics affecting geomorphic and hydrologic processes. Landform analysis with respect to their nature, extent, spatial distribution, potential and limitations is very useful for evaluation and optimal utilization of land resources. In quantitative approach for landform mapping and classification based on high resolution DEMs at watershed level, slope, elevation range, contours and stream network pattern can be used as basic identifying parameters. High-resolution satellite data provides reliable source of information to prepare comprehensive and detail inventory of landforms (Reddy and Maji, 2003), soil resources mapping, identification of groundwater potential zones (Reddy et al., 1999), landscape ecological studies (Reddy et al., 2001) and land degradation mapping (Reddy et al., 2004a). The surface lithological analysis of terrain can be carried out through interpretation of the satellite data in conjunction with collateral data. Visual interpretation techniques can be followed in delineation of landform units based on the characteristics of the satellite imagery in conjunction with drainage and contour information. The boundary delineation of landform units can be carried out based on the changes in topographic slopes, relief patterns, crest type, drainage texture and image characteristics of the area. Subsequently, detailed landform analysis can be carried out with the help of relief, morphometry and image characteristics at watershed level.

7.2.4 SOIL RESOURCE INVENTORY

Soil survey provides an accurate and scientific inventory of different soils, their kind and nature, and extent of distribution so that one can make prediction about their characters and potentialities (Manchanda et al., 2002). Landform units have been used as basic landform descriptors in soil resource mapping (Speight, 1990; Milne et al., 1995). Several authors have reported that the satellite remote sensing and GIS are proved as promising tools in large-scale soil resource mapping (Butte, 1997; Maji et al., 2004; Srivastava and Saxena, 2004; Sarkar et al., 2006). GIS application in soil survey and mapping can be divided into two parts. One is GIS applications in pre survey and mapping and another is the post survey and mapping applications. Firstly, a detailed landform map, derived from topographical sheets and satellite data analysis through visual interpretation, can be prepared to form a base map in soil survey, which shows spatial variation in terrain features and their characteristics. Secondly, based on the interpretation, the potentials and limitations of the soils can be obtained and such information could be used to construct geospatial database using GIS. Using geospatial applications, soil resource inventory has been conducted in Borgaon Manju watershed in semiarid AER (Akola district) and the results show nine distinct landform units and 14 soils series. The detailed description on soils will be of immense help in crop planning and management of soil resources towards land use planning and food security at watershed level (Reddy et al., 2008).

7.2.5 LAND USE SYSTEMS ANALYSIS

The size of land holding plays a key role in efficient management of land resources to accomplish sustained productivity, food security and environment. Study conducted in Borgaon Manju watershed shows that major crops grown during Kharif season are cotton, pigeonpea, sorghum, soybean and Bengal gram. During Rabi season wheat, Bengal gram, safflower and sorghum are the main crops. The analysis shows that in Borgaon Manju watershed, the single cropped area has been reduced by 8.0% over a period of 25 years (1980–1981 to 2005–2006). At the same time the double-cropped area was increased by 36.0%. Over a period of 25 years, the net cultivable area in the watershed was reduced by 4.4%. Similarly the gross cropped area also reduced by 1.2%. However, during the same period the cropping intensity was increased by 4.0% due to increased area under double crop. The reduction of area under net cultivable area might be due to urbanization/industrialization in the western part of the watershed. During the same period, the analysis shows that area under cotton/pigeon pea-fallow system has been reduced from 33.6 to 23.6% in the watershed. During the same period, area under cotton-fallow has been increased 2.7 to 9.0% of TGA of watershed (Reddy et al., 2008). As a case study, the temporal changes of land use systems in Dhotardi village of Borgaon Manju watershed for the year 1980–1981 and 2005–2006 are shown in Fig. 7.3.

FIGURE 7.3 Land Use Systems of Dhotardi Village in Borgaon Manju Watershed for the Year 1980–2081 and 2005–2006.

7.2   GEOSPATIAL TECHNOLOGIES IN LAND RESOURCE EVALUATION AND MANAGEMENT

The land resource evaluation forms a prerequisite for sustainable land use planning. Land use planning aims to encourage and assist land users in selecting options that increase their productivity, are sustainable and meet the needs of society (FAO, 1993). Geospatial technologies in soil-site suitability evaluation, soil loss assessment, watershed prioritization and carrying capacity assessment have been discussed in the following subsections.

7.2.1   SOIL SITE-SUITABILITY EVALUATION

Land use planning aims to encourage and assist land users in selecting options that increase their productivity, are sustainable and meet the needs of society (FAO, 1993). Land suitability evaluation is an examination process of the degree of land suitability for a specific utilization type (Sys et al., 1991). Since land suitability analysis requires the use of different kinds of data and information (soil, climate, land use, topography, etc.), the GIS offers a flexible and powerful tool than conventional data processing systems, as it provides a means of taking large volumes of different kinds of datasets and combining the datasets into new datasets (Foote and Lynch, 1996). With the assistance of a GIS, Liu and Deng (2001) developed a land resources management system to evaluate land suitability. Matching of the land attributes with the specific crop growth requirements and definition of the preliminary suitability classes was workout. Maji et al., (2005) followed multicriteria based decision making procedure to generate suitability maps for Bilaspur district of Chhattisgarh state. The thematic database on slope, soil depth, soil erosion, soil drainage, soil texture, surface stoniness, coarse fragments, $CaCo_3$ and OC layers were used in AGROMA GIS to find out the suitability areas for paddy. Walke et al. (2011) made an attempt to evaluate the soil resources of the Ringnabodi watershed, Nagpur district of Maharashtra, Central India, in soil-site suitability evaluation for cotton using multicriteria overlay analysis techniques in GIS.

7.2.2   SOIL LOSS ASSESSMENT

Soil erosion is a major concern in landscape management and conservation planning (FAO, 1985). Geospatial technologies play a critical in the generation of spatial data layers and their integration to estimate soil loss at watershed level by adopting the suitable models like Universal Soil Loss Equation (Wischmeier and Smith, 1978). GIS techniques have been used extensively in quantification of USLE parameters to estimate soil erosion (Jain and Kothari, 2000). The factors of rainfall erosivity (R), soil erodibility (K), slope length (LS), cover (C) and management (P) are the components of USLE and can be computed in GIS using rainfall data, series-wise soil information, slope generated from contours, land use/land cover interpreted from remotely sensed data and field data, respectively. GIS helps in integrated analysis of USLE parameters in quantification of potential and actual soil loss (Reddy, et al., 2004b). Following multicriteria overlay of several critical parameters the units for soil conservation can be delineated to suggest suitable agronomic and mechanical protection measures based on the soil and site characteristics in each of the delineated units.

7.2.3   WATERSHEDS PRIORITIZATION

Remote sensing and GIS techniques have been used extensively in quantification of USLE (R, K, LS, C and P) parameters for soil loss assessment (Biswas et al., 1999; Pandey et al., 2007; Yoshino and Ishioka, 2005). These layers can be integrated in GIS to quantify potential and actual soil loss. The study conducted in Vena watershed of Nagpur district by Reddy et al., 2002, 2004b shows that computation of morphometric parameters and USLE parameters at watershed level in GIS has been found very useful in evaluation of erosion characteristics for their prioritization. The comparative analysis of ranks obtained from morphometric, estimated potential and actual soil loss parameters showed good interrelationships and the ranks of morphometric parameters were considered to accretion the validity of ranks obtained from USLE parameters. Further, the erosion susceptibility zone map generated by grouping the watersheds based on their ranks has been used in identification of priority zones for the evaluation and suggestion of soil conservation measures. The study demonstrated that GIS and remote sensing approach in prioritization of watersheds and erosion susceptibility zone mapping based on ranks obtained from morphometric and USLE parameters is found to be more appropriate. The integrated remotely sensed data and GIS based approach also helps in identification of feasibility sites for in-situ soil and water conservation for optimum utilization of available land and water resources with minimum hazard to geo-ecological balance at watershed level.

7.2.4   CARRYING CAPACITY ASSESSMENT

The concept of sustainable development and food security are closely linked to the carrying capacity of ecosystems (FAO, 1992). The status of carrying capacity has been considered as one of the parameters in assessment of ecological sustainability at watershed level. Carrying capacity may be seen as the ability to produce desired outputs (goods and services) from limited resources while maintaining desired stock and quality of the resource base. Using GIS, the spatial variability of carrying capacity has been assessed in Borgaon Manju watershed, Akola district based on the availability and requirement of food grains in terms of calories for human population and availability and requirement of forage (dry matter) for animal population for the year 2005–06. The analysis shows that majority of villages in the watershed have Food grains Security Index (FSI) less than 100, indicating deficiency in meeting the food grains requirement. Further, it shows that out of the 37 villages in the watershed, about seven villages, which have the FSI rank '7' are highly deficit in food grains availability ranging from 50 to 75% over the requirement. Two villages, have the FSI rank of '8' are under extremely low in food grains availability of more than 75% over the requirement. Analysis of DSI in the watershed shows 20 villages have the DSI rank 8 and are under the category of extremely deficit in dry matter availability, which is about more than 75% of the requirement. The analysis of composite carrying capacity index (CCI) shows that very low CCI zone with the rank 7 have very low in meeting the food grains demand and fodder requirement and it ranges from 50 to 75% deficit. The analysis indicates that the advance capabilities of GIS have immense potential to understand the spatial variability of carrying capacity for sustainable management of natural resources and food security (Reddy et al., 2008).

7.2.5 SPATIAL DECISION-SUPPORT SYSTEM (DSS)

DSS is explicitly designed to support a decision-making process for complex spatial problems. SDSS provides a framework for integrating Database Management System (DBMS) with analytical models, graphic display, and tabular reporting capabilities and expert knowledge of decision makers. The modeling capability of GIS allows the user to develop SDSS to generate criterion-based scenarios in different climatic and terrain conditions. The database components of the SDSS can supply input data for the models and the resulting output can be returned to the database for later display through user interface, in tabular, chart or map form. This GIS based SDSS helps in modeling capability with simulation techniques to solve complex land resource management problems. DSS is an interactive computer-based system or subsystem intended to help decision makers to identify and solve problems, complete decision process tasks, and make decisions. By developing comprehensive database through DSS, the problems like land resource management and area specific land use scenarios with better livelihood options could be developed to ensure optimum utilization of land resources and food security at watershed level.

7.3 CONCLUSION

The use of modern geospatial technologies such as high resolution satellite data, GPS and GIS can be effectively used in land resources inventory, mapping, monitoring and management. The satellite data provides integrated information on landforms, geological structures, soil types, erosion, land use/land cover, surface water bodies and qualitative assessment of groundwater potentials at watershed level. Geospatial technologies can be effectively used in generation of digital terrain database, inventory and mapping of landforms, soils, land use systems, soil suitability evaluation, soil loss assessment, prioritization of watershed and assessment of carrying capacity at watershed level. The generated precise database in GIS on land resources, socioeconomic and contemporary technologies and integrated analysis provide a sophisticated mechanism to generate alternative action plans for watershed development. These action plans can then be presented to the concerned people for sensitizing them to get involved and to integrate their planning with reflections of their aspirations with a convergence approach in decision making to improve the productivity levels to meet ever increasing food, fodder and fuel demand as well as to ensure food security.

KEYWORDS

- **Borgaon Manju Watershed**
- **Carrying Capacity Index**
- **Database Management System**
- **Digital Elevation Model**
- **Digital Terrain Database**
- **Digital Terrain Database Generation**
- **Digital Terrain Modeling**
- **Food grains Security Index**
- **Geographic Information System**
- **Geospatial Technology**
- **GIS Technology**
- **Global Positioning System**
- **Land Use Systems Analysis**
- **Shuttle Radar Topographic Mission**
- **Soil Resource Inventory**
- **Spatial Decision-Support System**

REFERENCES

Bertolo, F. (2000). Catchment delineation and characterization: A review, Tech. Rep. EUR 19563 EN, Joint Res. Cent. Eur. Comm., Ispra, Italy.

Biswas, S., Sudhakar, S., Desai, V. R. (1999). Prioritization of sub watersheds based on Morphometric analysis and drainage basin: A remote sensing and GIS approach, Journal of Indian Society of Remote sensing Vol.27, No.3: 155–166.

Butte, P. S. (1997). Application of IRS-IC satellite data in land resource appraisal and planning of Mohgaon subwatershed, Nagpur district, M. S. M.Sc (Agril.) Thesis (Unpubl.) Dr. PDKV, Akola (M.S).

FAO, (1985). Watershed development with special reference to soil and water conservation, Soil Bulletin 44, Food and Agricultural Organization, Rome, 1985.

FAO, (1992). Technical report, Food and Agriculture Organization of the United Nations.

FAO, (1993). Guidelines for Land Use Planning. FAO Development Series 1, FAO/AGLS, Rome, 96 p.

Foote, K. E., Lynch, M. (1996). Geographic Information Systems as an Integrating Technology: Context, Concepts and Definition. Vol. 2. University of Texas, Austin, USA.

Frankel, K. L., Dolan, J. F. (2007). Characterising arid region alluvial fan surface roughness with airborne laser swath mapping digital topographic data. Journal of Geophysical Research, 112, F02025.

Jain, M. K., Kothyari, U. C. (2000). Estimation of soil erosion and sediment yield using GIS, Hydrological Sciences Journal, 45 (5),771–786.

Krishnamurthy, J., Srinivas, G. (1995). Role of geological and geomorphological factors in groundwater exploration: A study using IRS-LISS-II data. International Journal of Remote Sensing, 16, 2595–2618.

Li, Z., Zhu, Q., Gold, C. (2005). Digital Terrain Modeling: Principles and Methodology. CRC Press, Boca Raton.

Liu, X., Zhang, Z. (2010). Drainage network extraction using LiDAR-derived DEM in volcanic plains. Area, doi: 10.1111/j.1475–4762.2010.00955.x.

Liu, Y. S., Deng, X. Z. (2001). Structural patterns of land types and optimal allocation of land use in Qinling Mountainous. Journal of Geographical Sciences, 11(1), 99–109.

Maji, A. K., Reddy, G. P. O., Tamgadge, D. B. Gajbhiye, K. S. (2005). Spatial Modeling for Crop Suitability Analysis Using AGROMA GIS software, Asian Journal of Geoinformatics, Vol. 5 (3), 47–56.

Maji, A. K., Reddy, G. P. O., Thayalan, S., Walke, N. J. (2004). Characterization and Classification of Landforms and Soils over Basaltic Terrain in Sub-humid Tropics of Central India, Journal of Indian Society of Soil Science, Vol 53 (2),154–162.

Manchanda, M. L., Kudrat, M., Tiwari, A. K. (2002). Soil survey and mapping using remote sensing, Tropical Ecology 43(1), 61–74.

McBratney, A. B., Santos, M. L. M., Minasny, B. (2003). On digital soil mapping. Geoderma, 117(1–2), 3–52.

Milne, J. D. G., Clayden, B., Singleton, P. L., Wilson, A. D. (1995). Soil Description Handbook. Manaaki Whenua Press, Landcare.

Moore, I. D., Gessler, P. E., Nielson, G. A. (1993). Soil attribute prediction using terrain analysis. Soil Science Society of America Journal, 57, 443–452.

Moore, I. D., Grayson R. B. Ladson, A. R. (1991). Digital Terrain Modelling: A Review of Hydrological, Geomorphological, and Biological Applications, Hydrological Processes, 5(1), 3–30.

Paik, K. (2008). Global search algorithm for nondispersive flow path. Journal of Geographical Research, 113:doi:10.1029/2007JF000964.

Pandey, A., Chowdary, V. M., Mal, B. C. (2007). Identification of critical erosion prone areas in the small agricultural watershed using USLE, GIS and remote sensing, J. Water Resource Management, 21: 729–746.

Pike, R. J., Acevedo, W., Thelin, G. P. (1988). Some topographic ingredients of a geographic information system, Proceedings, International Geographic Information Systems Symposium, 15–18 November, Arlington, Virginia (NASA, Washington, D. C.), 2~151–164.

Reddy G. P. O., Maji A. K, Srinivas C. V., Thayalan, S., Velayutham, M. (2001). Landscape Ecological Planning in a Basaltic Terrain, Central India, Using Remote sensing and GIS Techniques, Journal of Indian Society of Remote sensing, Vol.29 (1&2), 3–16.

Reddy, G. P. O, Chandramouli, K., Srivastav, S. K., Maji, A. K., Srinivas, C. (1999). Evaluation of groundwater potential zones using Remote sensing data – A case study. Journal of Indian Society Remote Sensing, Vol.28, (1), 19–32.

Reddy, G. P. O., Jagdish Parsad, Ramamurthy, V. Maji, A. K., Gajbhiye, K. S. (2006). Development of Resource Database and the Spatial Modeling for Integrated Development of Gondia District, Maharashtra, Project Report, NBSS & LUP, Nagpur. 50 p.

Reddy, G. P. O., Maji, A. K. (2003). Delineation and Characterization of Geomorpholo-gical features in a part of Lower Maharahstra Metamorphic Plateau, using IRS-ID LISS-III data, Journal of the Indian Society of Remote sensing, Vol.31 (4), 241–250.

Reddy, G. P. O., Maji, A. K., Chary, G. R., Srinivas, C. V., Tiwary, P., Gajbhiye, K. S. (2004b). GIS and Remote sensing Applications in Prioritization of River sub basins using Morphometric and USLE Parameters – A Case study. Asian Journal of Geoinformatics, Vol.4 (4), 35–49.

Reddy, G. P. O., Maji, A. K., Gajbhiye K. S. (2004a). Drainage Morphometry and Its Influence on Landform Characteristics in Basaltic Terrain – A Remote Sensing and GIS Approach. International Journal of Applied Earth Observation and Geoinformatics, Vol.6 (1),1–16.

Reddy, G. P. O., Maji, A. K., Nagaraju, M. S. S., Thayalan, S., Ramamurthy, V. (2008). Ecological evaluation of land resources and land use systems for sustainable development at watershed level in different agro-ecological zones of Vidarbha region, Maharashtra using Remote sensing and GIS techniques, Project Report, p. 270, NBSS&LUP, Nagpur, India.

Reddy, G. P. O., Maji, A. K., Srivastava, R., Das, S. N. (2011). Development of GIS based Seamless Mosaic of SRTM Elevation Data for India to analyze and characterize the selected geomorphometric parameters, Project Report, NBSS&LUP, Nagpur, 1–54.

Reddy, G. P. O., Maji., A. K., Chary, G. R., Srinivas, C. V., Tiwary, P., Gajbhiye, K. S. (2002). Geo-spatial database for prioritization and evaluation of conservation strategies for landscape management in a river basin, Proceedings of ISPRS Commission VII, Symposium on Resources and Environmental Monitoring, Hyderabad, India, (Ed. R. R. Navalgund et al.), Vol.1, 654–659.

Sarkar, D., Gangopadhyay, S. K., Sahoo, A. K. (2006). Soil resource appraisal towards land use planning using satellite remote sensing and GIS- A Case study in Patloinala microwatershed, District Puruliya, West Bengal, Journal of the Indian Society of Remote sensing, 34 (3), 245–260.

Speight, J. G. (1990). Australian Soil and Land Survey: Field Handbook, 2nd ed. Inkata Press, Melbourne.

Srivastava, R., Saxena, R. K. (2004). Techniques of large-scale soil mapping in basaltic terrain using satellite remote sensing data, International journal of remote sensing, 25(4), 679–688.

Sys, C., Vanranst, E., Debvay, J. (1991). Land Evaluation, Part: III. General Adminstration for Development Coopration Agriculture, Brussels, Belgium.

Tribe, A. (1992). Automated recognition of valley lines and drainage networks from grid digital elevation models: a review and new method. Journal of Hydrology, 139: 263–293.

Walke, N., Reddy, G. P. O., Maji, A. K., Thayalan, S. (2012). GIS-based multicriteria overlay analysis in soil-suitability evaluation for cotton (Gossypium spp.), A case study in the black soil region of Central India, Computers and Geosciences, Vol. 41: 108–118.

Westman, W. (1985). Ecology, Impact Assessment and Environmental Planning. New York: Wiley.

Wilson, D. J., Gallant, J. C. (2000a). Digital terrain analysis. In: D. J. Wilson and J. C. Gallant (Editors), Terrain Analysis: Principles and Applications. John Willey & Sons, INC, New York, pp. 1–27.

Wilson, P. J., Gallant, J. C. (2000b). Secondary topographic attributes. In: P. J. Wilson and J. C. Gallant (Editors), Terrain Analysis: Principles and Applications. John Willey & Sons, INC, New York, pp. 87–131.

Wischmeier, W. H., Smith, D. D. (1978). Predicting rainfall erosion losses – A guide to conservation planning, USDA Agricultural Handbook No. 587.

Yoshino, K., Ishioka, Y. (2005). Guidelines for soil conservation towards integrated basin management for sustainable development: a new approach based on the assessment of soil loss risk using remote sensing and GIS, Paddy. Water. Environ., 3, 235–247.

Young, A. (1998). Land resource: Now and for the Future, Cambridge, Cambridge University Press.

CHAPTER 8

KNOWLEDGE MANAGEMENT INITIATIVE THROUGH AGRICULTURE SECTOR THROUGH AGRO CLIMATIC PLANNING AND INFORMATION BANK FOR LAND RESOURCES DEVELOPMENT

V. MADHAVA RAO[1], T. PHANIDRA KUMAR[2], R. R. HERMON[3], P. KESAVA RAO[4], S. S. RAVI SHANKAR[4], and N. S. R. PRASAD[4]

CONTENTS

[1]Professor and Head.

[2]Project Consultant, Centre on Geoinformatics Application in Rural Development (CGARD), National Institute of Rural Development and Panchayati Raj (NIRD & PR), Ministry of Rural Development, Government of India, Hyderabad

[3]Associate Professor

[4]Assistant Professor

ABSTRACT

Agro Climatic Planning and Information Bank (APIB) is a multipronged approach to develop databases to contain information on both Spatial and Non-Spatial parameters, with a Decision Support System (DSS) providing a frame work for integrating databases management systems, analytical models and graphics in order to improve decision-making process, in agriculture up to plot and farmer level, addressing the last mile issues. APIB covers agricultural land use/land cover; cropping systems; irrigated crop land; soils; degraded land; geomorphology; ground water potential; drainage characteristics cultural; and features: agro-ecological characterization, database for Agricultural Practices, Seeds, Fertilizers, Plant Protection, Agricultural Implements, Agricultural Credits, Insurance Schemes, Subsidy, Agriculture Market Infrastructure, Weather Information, Socio-economic Infrastructure, sustainable agricultural land use plan based on integration of land capability, land productivity, soil suitability, terrain characteristics and socioeconomic information using geospatial technology. APIB is a Knowledge Management in Agriculture Practice, making economically viable, productivity and growth oriented and integrating with the enrichment in the progress and livelihood of farmers and rural agro economy.

8.1   INTRODUCTION

Many developing countries in the South Asia, like India, with a vast geographical area are bestowed with the bounties of natural resources, namely minerals, soils, water, flora and fauna, and marine resources. Over exploitation of available natural resources for meeting the increasing demand for food, fuel and fiber of ever-growing population has led to degradation of land by way of soil erosion by water and wind, salinization and alkalinization, water logging, shifting cultivation, etc. An estimated 175 m ha of land are subject to some kind of degradation. Soil erosion by water and wind alone accounts for an estimated 150 m ha. In addition, water logging, soil salinization and alkalinization, and shifting cultivation have affected an estimated 6 m ha, 7.16 m ha and 4.36 m ha of land, respectively. Degradation of land by deforestation, forest fire, frequent floods and drought, further compounds the problem. For optimal utilization of available natural resources and for taking up any preventive or curative measures, timely and reliable information on natural resources with respect to their nature, extent and spatial distribution; and nature, magnitude and temporal behavior of various types of degraded lands, is a prerequisite. Hitherto, such information has been generated through conventional approach using topographical sheets or aerial photographs. Synoptic view of a fairly large area provided by multispectral measurements made from satellite platforms at regular intervals enable generating information on natural resources, degraded lands and environment in a timely and cost- effective manner.

Agriculture constitutes the backbone of the Indian economy. With rapidly rising population, already a billion plus, threatening to make India the most populated country in the world in next few decades, one has to look towards advanced technologies to help in meeting the needs of this burgeoning population for food, fiber and fuel. There is also need for utilization of our natural resources in a sustainable manner. Remote sensing and Geomatics technologies have demonstrated the potential for assisting in the management of these precious resources. At the international level, ever since the launch of Landsat 1, the first civilian remote sensing satellite in July 1972, the indispensable role of satellite remote sensing in agriculture sector has been fully appreciated. In fact, the choice of spectral bands in most of the civilian remote sensing satellites launched by various countries, including India, to- date has been dictated by their use for efficacious management of resources related to agriculture. This capability has been further enhanced manifold by Geomatics during the last few years, especially with the availability of high-resolution satellite images from IRS and Cartosat satellites.

Geomatics deals with the tools/techniques related to the measurement of geographical data, the data that has both spatial and nonspatial components. Most common among these tools are remote sensing, geographical information system (GIS) and Global Positioning System (GPS). While remote sensing data provides a synoptic, multispectral and repetitive information about any resource, the GIS helps in integrating the various spatial thematic layers (including remote sensing data) to arrive at an integrated decision. GPS enables the user to know the precise location of a target, which may be useful in identification, mapping or precise operation. Agriculture is a resource, which has got a large geographical extent. Especially in a country like India, where agriculture is the backbone of the country, there is great need of getting spatial information about this sector. The green revolution of 1960's has made the country's food production progress remarkable. The requirement has grown from less than 50 million tons in 1950's to more than 200 million tons at present. However, green revolution is also associated with many negative implications including the environmental degradation. Hence, in the 20-first century, to steer the agricultural achievements towards the path of an 'evergreen revolution,' there is a need to blend the traditional knowledge with frontier technologies. Information and communication technology; space technology; GIS are the tools of such frontier technologies which would help in creating agricultural management systems; making plans for sustainable agriculture; and bringing new areas (through development of wastelands) into productive agriculture. This present article tries to explore the role of Geoinformatics with special emphasis on remote sensing and GIS in agricultural development.

The role of remote sensing and GIS in agricultural applications can be broadly categorized into two groups – inventorying/mapping and management. While remote sensing data alone is mostly used for, inventorying (crop acreage estimation, crop condition assessment, crop yield forecasting, soil mapping, etc.) purposes, the management (irrigation management, cropping system analysis, precision farming, etc.) needs various other types of spatio physical environmental information. The latter has to be integrated with remote sensing data, where the functionality of GIS is used effectively for governance and for decision support.

The application of Geospatial technology in India in various sectors namely in wasteland and watershed management, water resources, irrigation, drinking water, health, infrastructure, road network, agriculture and other sectors of development, has brought in visibility, knowledge and scientific decision making at all levels, in a time and cost saving manner.

8.2   AGRO CLIMATIC PLANNING AND INFORMATION BANK (APIB)

Agriculture is still the primary sector in Indian economy. While, Green revolution's food production increased from about 5.1 million tones in 1950 to over 205 million tones now, the higher economic growth and population pressure demands about 260 million tones of food grains by 2030. It is therefore imperative that an accelerated pace of food production could be possible with modern technologies, information technology, space technology and judicious use of natural resources.

Further critical aspects of agriculture like seeds, financial support, fertilizer, marketing support, agro-processing and post harvest technologies, etc., are expected to improve the productivity of and profitability from Indian agriculture.

Agriculture reform is the need of the hour to bring better productivity through utilization of new technology. APIB initiated in 1998, initially by ISRO, at 1:50,000, on a pilot basis in Karnataka and subsequently in Meghalaya, for one district each, was seen as a promise by planners, nurtured by technologists and implemented by farmers. APIB with its concept of 'single window' access to knowledge base in the field of agriculture sector is eminently suitable to facilitate the reform process.

One of the important facilitating factors to enable more effective and participatory planning and decision making will be to create an access to what may be called knowledge base which has bearing on decision making and activities of various stakeholders/users.

The final decision in agriculture rests with the cultivator. Given the local agroclimatic and land-endowments, the farmer's decisions are largely influenced by factors like availability of water, inputs, credit and overall infrastructure situation including the access to the market. Yet he has a variety of options and he exercises his choice to the best of his knowledge and ability. Hence, the farmer is in need of being aided as to what are his best options by access to a comprehensive land use and management plans and assessment of potential, which represents these options.

This activity will not be limited to hand ling over the plan printed on paper but also providing with back-up information and expert advice and should be constantly updated in the light of changes in the environment and technology to fulfill the requirements of sustainable agricultural land use planning.

APIB requirement is multipronged approach to develop databases to contain information on both Spatial and Non-Spatial parameters. The linkages between non-spatial and spatial databases will be facilitated through GIS. DSS provide a framework for integrating databases management systems, analytical models and graphics in order to improve decision-making process.

For Crop Planning based on watershed as a unit GIS is an ideal tool since Digital Elevation Model (DEM) created in GIS can be integrated with other spatial information including land use pattern obtained from Remote Sensing Data from time to time is very important for optimizing the use of limited water resources for maximizing and sustaining the productivity under rainfed conditions.

This vision as reflected in Precision Farming aims to direct the application of Seed, Fertilizers, Pesticides and Water within fields in ways that optimize farm returns and minimize pesticides and environmental hazards. Adoption of this technology requires accurate natural resources maps showing physical and chemical properties of the soils and the tools to apply the inputs as per the spatial variability.

Thus APIB is emerging as Knowledge Management in Agriculture Practice, making economically viable, productivity and growth oriented process and integrating with the enrichment in the progress and livelihood of common men.

8.3   APIB OBJECTIVES

The APIB primarily aims at addressing the following objectives:
1.   To prepare agriculture and natural resources inventories by analysis of temporal satellite data, including agricultural land use/land cover, cropping systems, irrigated crop land, soils, degraded land, geomorphology, ground water potential, drainage characteristics, cultural features, etc.
2.   To collect agro-climatic and agricultural data for agro-ecological characterization and agro-climatic regional planning.
3.   To develop a database for Agricultural Practices, Seeds, Fertilizers, Plant Protection, Agricultural Implements, Agricultural Credits, Insurance Schemes, Subsidy, Agriculture Market Infrastructure, Weather Information, Socio-economic Infrastructure, etc., in GIS.
4.   Development of Spatial Decision Support System (SDSS) for agro-climatic planning and information bank for agricultural reform process as well as the management of agricultural land and other agricultural allied activities.
5.   Suggesting sustainable agricultural land use plan based on integration of land capability, land productivity, soil suitability, terrain characteristics and socioeconomic information, etc., using GIS.

8.4   PILOT STUDY AREA

The study was demonstrated in two districts of Champawat and Dehradun District in the State of Uttarakhand, in the northern part of India and envisaged to be subsequently extended to other districts of the State (Fig. 8.1). The information bank extends up to village level with soil profiling and mapping up to micro level applications, useful for decision making in agriculture sector, linking all related sectors of rural economy.

To carry out detailed study using ground truthing, maps, field observation, high resolution IRS multispectral and multitemporal images in GIS environment, in 1:10,000 scale large scale mapping for agriculture areas and 1:50,000 scale maps for forest and steep hilly areas, the study area chosen was the Champawat and Dehradun districts of Uttarakhand State, All spatial features are being carried out with a detailed study in the District of Champawat in Uttarakhand state at 1:10,000 scale, for agricultural areas and forestry/hilly terrain at 1:50,000 scale.

FIGURE 8.1   Study Area Maps of Champawat and Dehradun Districts.

8.5   PROCESS METHODOLOGY

APIB is expected to play an important role in the agricultural reform process and in developing agricultural planning and development. It established linkages with various organizations, efforts made in synthesizing of data, dissemination of planning advise and information modules developed for serving the diverse user clientele in the agriculture sector. The 'process methodology' integrates all aspects of agroclimatic and natural resources database.

8.6   AGRO-CLIMATIC AND NATURAL RESOURCES DATABASE

In order to achieve the various objectives envisaged under APIB, an assessment of its clientele's planning needs in the field of agriculture, viz., farmers, rural financial institutions, input/output traders, agricultural researchers, extension workers, planners, etc., are required to be studied and analyzed. Based on the needs expressed by the user clientele, APIB has followed a multipronged approach to develop databases. As a first step, APIB has designed a database structure to contain information on both spatial and nonspatial elements. For example, in the nonspatial domain, APIB information system contains modules on fertilizers, plant protection, seeds/seedlings, package of practices, agricultural implements, climatological data, credit/insurance schemes available, infrastructure for processing and marketing, demographic details, etc. Similarly, in the spatial domain, it has information on various resources like land use, groundwater, soils, slope, etc.

(a) Non- spatial Database: Given below is a brief summary of APIB nonspatial database, which shall be generated for the two districts of Uttarakhand State:

(i) Package of Practices Module: Package of practices collected for all major crops like cereals, pulses, cash crops; oil seeds, fodder crops, fruit crops, spices, vegetables, floriculture, medicinal and aromatic plants, etc. Package of practices are also available for many high-yielding hybrids produced by many private companies. Value addition in the form of recent market prices, support prices, inputs suppliers, etc., are being developed.

(ii) Seeds/seedlings Module: This will contain information on the special characteristics/salient features including information like yield, resistance to pests/diseases, quantity of seeds required, crop duration, crop suitability for different regions, packing and price information and dealer network in the priority districts.

(iii) Fertilizers Module: The module will have detailed information on type of fertilizers, and their technical formula, commercial trade names, prices and dealers network in the priority districts. The software package for recommending fertilizer dosage based on soil test and crop type shall be developed.

(iv) Plant Protection Module: Information content in this module will include the type of chemicals, their active ingredients, trade names, prices, applicability to different crops/pests, dealers/suppliers, etc. This module will also contain information regarding biological pest control as well as other conventional pest control agents. In addition to the information on integrated pest management practices for selected crops and the plant protection guide for horticultural crops.

(v) Agricultural Implements: Under this module, APIB has information on various hand tools, bullock/tractor drawn implements, irrigation equipment, plant protection equipment, harvest and postharvest equipment, their suppliers, price information, models, salient features, etc.

(vi) Agricultural Credit: Information readily available with the Banks will cover different credit schemes in agriculture and allied sectors, floated by different commercial rural banks in the districts. The details of each scheme covering information on interest rates, repayment terms, documents to be produced and other terms and conditions of schemes, location addresses of the branches of the different banks, specimen copies of applications (available for a few banks), etc., will also be available.

(vii) Insurance for Agriculture and Allied Sectors: This module has details of various insurance schemes for crops, plantation, agricultural implements and farm machinery, animal husbandry, including information on terms and conditions of the schemes, premium rates, eligibility criteria/procedure and benefits as well as the branch network of the insurance companies.

(viii) Subsidy Programs for Agriculture: The database under this module contains vast information on the various subsidy programs available for farming sector beneficiaries in the four priority districts of Uttarakhand. Information on actual availability procedure with details of eligibility criteria, subsidy, credit components, documentation, etc., is available.

(ix) Agriculture Market Infrastructure: APIB has synthesized information on market infrastructure facilities in its four priority districts of Uttarakhand. This data module provides information on the network of Agriculture Produce Marketing Committees (APMC) in these districts, market charges, godown and storage facilities available and their charges. APIB will have information on export policies applicable for selected farm products. The price information for selected commodities shall also be available from Uttarakhand State Agricultural Marketing Board with whom APIB will establish a formal institutional contact for market related data exchange.

(x) Agricultural Weather Information System: Weather is a very critical parameter for agriculture and allied sectors. Most of the farming activities are guided by weather conditions in an area. APIB shall develop a weather information system. Based on weekly and monthly rainfall data/moving averages for rainfall and weekly probabilities of rainfall, long-term climate change and trends in rainfall patterns shall be identified for these districts.

(xi) Training and Development: The module has the details on types of training, periods of training, benefits as well as the locations of farmers' training centers.

(xii) Socio-Economic Infrastructure: The village-wise census details on demography and amenities have been collected. In addition, also block-wise infrastructural details were collected. Computerization of basic data addresses for most of the information needs, accomplished. Additionally all the relevant socioeconomic information for at least the past five years is also made available.

(b) Spatial Database

The strength of APIB is a powerful state-of-the-art image analysis and GIS facilities to assess the natural resources endowments of any agro-climatic region in

Uttarakhand, in spatial and temporal domain, for taking decision by farmers at plot level.

Data Products used were high resolution Satellite Imageries that of Cartosat-I and Resource Sat P5 LISS-IV and associated Topo Sheets from Survey of India at 1:25,000 scale. Detailed Soil Survey were carried on at an interval of 200–500 meters interval based on soil variability covering all villages, of the Champawat and Dehradun Districts, for agricultural area. Spatial agro-meteorological Characteristics Rainfall, Air temperature (Minimum and Maximum), Solar radiation (if-available) monthly basis for past 30 year were collected from Indian Meteorological Department, Pune. Socio-Economic Data Village wise, social group wise and gender data from Census were procured from Census Department and integrated. Revenue Maps were collected from Department of Agriculture, covering up to cadaster. Further GPS/DGPS data were collected to augment these maps for necessary corrections for reliability. These digital layers were integrated with the spatial data layers generated for the Champawat and Dehradun Districts, subsequently. Detailed parameters considered for the study are presented in Table 8.1. Further location specific Packages of Agronomic Practices in vogue provided by the Department of Agriculture, were also integrated in the APIB System.

TABLE 8.1 List of Parameters Considered for Study

S. No.	Parameters	Source	Period	Utilization
1	Soil Mapping	Satellite and detailed field survey data	March/ April	Land use Planning; Soil management
2.	Dominant Crop Types	Satellite data	Rabi and Kharif	Agriculture Management, Crop production forecasting
3.	Landuse	Satellite data	Rabi and Kharif	Landuse planning
4.	Land degradation	Satellite data	March	Land reclamation, planning control measures
5	Geomorphology	Satellite data	March	Geological hazard assessment and Ground water potential
6.	Ground water potential	Derived from geomorphology and other terrain information	Summer Season Data	Water resources planning; irrigation planning

TABLE 8.1   *(Continued)*

S. No.	Parameters	Source	Period	Utilization
7.	Irrigated/unirri-gated land areas	Satellite data and irrigation command area information	Rabi and Kharif	Irrigation water management
8	Soil physico-chemical proper-ties	Field survey and Lab. Analysis of soil samples	Khaif season	Soil classification and Soil management
9	Terrain char-acteristics (el-evation, slope, aspect)	Derived from DEM (Toposheet)	Rabi and Kharif	Agricultural planning, agro ecological zonation
10	Agro-meteoro-logical data	Meteorological station	Daily/ weekly/ monthly	Agricultural planning; crop yield prediction
11	Cultural features (roads, canal, rail network, villages, towns etc.)	Satellite data, Topo sheet	Rabi and Kharif	Infrastructure planning
12	Drainage char-acteristics	Satellite data, Topo sheet	Rabi and Kharif	Watershed management
13	Soil loss	Soil map and soil properties data; climatic and terrain char-acteristics data; modeling	Rabi and Kharif	Soil erosion hazard assessment, soil conservation planning
14	Soil suitability for crops	Soil map and soil properties data; climatic and terrain char-acteristics data	Rabi and Kharif	Soil management, land use planning
15	Hydrological soil grouping	Soil map and soil properties data; climatic and terrain char-acteristics data	Rabi and Kharif	Watershed hydrology, run-off estimation

TABLE 8.1   *(Continued)*

S. No.	Parameters	Source	Period	Utilization
16	Land irrigability	Soil map and soil properties data; climatic and terrain	Rabi and Kharif	Irrigation development, irrigation water management
17	Land productivity	Soil map and soil properties data; climatic and terrain characteristics data	Rabi and Kharif	Soil management, land use planning
18	Land capability	Soil map and soil properties data; climatic and terrain characteristics data	Rabi and Kharif	Soil management, land use planning

APIB (Agro Climatic Planning and Information Bank) is a single window Knowledge base for Agricultural and allied Sector. It will help the Government and farmers to make decisions for improving agricultural productivity. APIB has designed a database structure to contain information on both spatial and non spatial elements namely mapping, GPS survey, soil survey, GIS and Image Processing analysis, fertilizers use, plant protection, seeds/seedlings, package of practices, agricultural implements, climatological data, credit/insurance schemes available, infrastructure for processing and marketing, demographic details, etc. The strength of APIB is a powerful state-of-the-art image analysis and GIS facilities to assess the natural resources endowments to any agro-climatic regions and would present them in the spatial and temporal domain. The Natural Resources databank are generated for the 18 themes at 1:10,000 scale. All the database are generated at 1:10,000 scale level, except in case of very steep and steep hilly areas under forest where the level of mapping will be reduced to 1:50,000 scale.

Through Village GIS, which is a customized GIS package for Village Information System, is possible to retrieves the micro level information like Household, Cadaster, MGNREGS, etc., of a village along with respective maps on the ground (Fig. 8.2). Apart from this it also displays Census and Infrastructure details of a village. All the information is linked to the maps and by clicking on the area of interest, all information is accessed (Figs. 8.3–8.5). This package is very much helpful to identify account holder, and category of the land, and area reserved for a particular purpose. This customized package has all analytical features and can handle audio, video, and photographs, etc. The information base has all Point, Line, Polygon features and Cadaster with Ownership details and Household information along with Photographs.

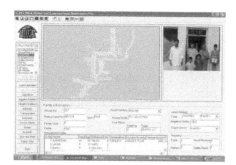

FIGURE 8.2   Gram Panchayat and Household Information Template.

Soil Map of the Study Area

Agriculture Suitability Map of the Study Area

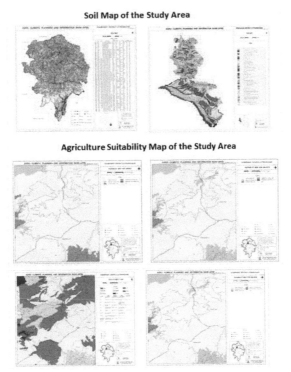

FIGURE 8.3 Study Area Maps Pertinent to Soils, Agriculture Suitability and Land Productivity.

For Apple   For Citrus

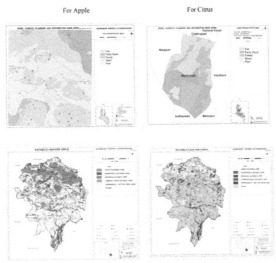

FIGURE 8.4 Land Suitability Maps for Apple and Citrus Plantations.

FIGURE 8.5   Cadastral Maps of the Study Villages.

8.7   CONCLUSION

India is at the forefront of Geoinformatics technology, while agriculture in India, is under serious stress. Modern technologies like GIS, Remote Sensing, GPS and other emerging technologies like Sensor based technologies, complementing agriculture technologies namely biotechnology, precision agriculture and modern irrigation technologies, is excepted to meet the growing demand of India in food security, by improving productivity in agriculture practice and optimum use of resources, which is emerging, as a knowledge management in agriculture sector.

GIS is a tool of power. Like all geographical information, GIS has the potential to reduce social inequities or to exacerbate them, within regions. GIS can be subversive or it can empower techno-elite (Clark and Worobec 1996). But Yapa (1991, 1952) argues that GIS may also be an instrument for 'discovering' local resources contextually and that the full implementation of appropriate technology is not possible without access to a GIS because it is the knowledge of the region (and the ability of the GIS to enhance this knowledge) that makes appropriate technology a viable alternative to the current modes of development.

In a global thinking and technology GIS could become a tool in the service of the poor rather than a technological instrument for their control. To that end, GIS and ICT and other spatial technology tools need to converge and emerge as powerful application tools for wider use at all levels, particularly at local or cadastral level, to take the benefit of technology to the door step of the poorest of the poor, to empower the people for making their lives better, livable and for leading a life with quality inputs for sustenance.

KEYWORDS

- **Agriculture Practice**
- **Agro Climatic Planning and Information Bank**
- **Cartosat Satellites**
- **Decision Support System**
- **Digital Elevation Model**
- **Geographical Information System**
- **Geoinformatics Technology**
- **Geomatics Technologies**
- **Global Positioning System**
- **Knowledge Management**
- **Process Methodology**
- **Remote Sensing**
- **Spatial Decision Support System**

REFERENCES

Anji Reddy. (2000). Remote Sensing and Geographical Information – An Introduction, Centre for Environment, Jawaharlal Nehru Technological University, Hyderabad.

Aronoff, S. (1989). Geographic Information Systems: A Management Perspective, WDL Publications, Ottawa, Canada.

Bhagavan, S. V. B. K. (1998). Remote Sensing Techniques for Watershed Programs in Andhra-Pradesh, Centre for Water Resources, J. N. Technological.

Gautham, N. C., Narayan, L. R. A. (1982). Suggested Land use/Land cover classification System for India using Remote Sensing Technique, Pink Publishing House, Mathura.

Gautham, N. C., Narayan, L. R. A., (1998). Wasteland Mapping in India, Pink Publishing House, Mathura.

NRSA, (1991). Guidelines to use Wasteland Maps. National Remote Sensing Agency, Hyderabad.

CHAPTER 9

DEVELOPMENT OF A WEB GIS BASED DECISION SUPPORT SYSTEM FOR AGRICULTURE CROP MONITORING SYSTEM

D. B. SURESH BABU[1], T. PHANINDRA KUMAR[2],
V. MADHAVA RAO[3], and I. V. MURALIKRISHNA[4]

CONTENTS

[1]COO and Director, Suchan Infotech Ltd, #296, Road No: 25, Jubileehilss, Hyderabad–500033

[2]Consultant, C-GARD, National Institute of Rural Development and Panchayati Raj (NIRD & PR), Hyderabad–500030

[3]Professor and Head, C-GARD, National Institute of Rural Development and Panchayati Raj (NIRD & PR), Hyderabad–500030

[4]Retired Prof and Director IST JNTU, Hyderabad–500030

ABSTRACT

The success of planning for developmental activities depends on the quality and quantity of information available on both natural and socioeconomic resources. It is, therefore, essential to devise the ways and means of organizing computerized information system. These systems must be capable of handling vast amount of data collected by modern techniques and produce up to date information. Remote Sensing technology has already demonstrated its capabilities to provide information on natural resources such as crop, land use, soils, forest, etc., on regular basis. The role of remote sensing and GIS in agricultural applications can be broadly categorized into two groups – inventorying/mapping and management. While remote sensing data alone is mostly used for, inventorying, crop acreage estimation, crop condition assessment, crop yield forecasting, soil mapping, etc., purposes, the management related activities like irrigation management, cropping system analysis, precision farming, etc., needs various other types of spatial physical and Environmental information. The latter has to be integrated with remote sensing data, where the functionality of GIS will be used. In the present study, techniques of remote sensing and geographic information system (GIS) have been used to monitor and map the Sugarcane crop at farm level. The Quick bird data of 60 cm resolution was acquired and was geometrically corrected and georeferenced for subsequent processing and analysis using Digital Image Processing (DIP) and GIS. After the generation of the thematic layers at farm level using GIS, using the various customization tools available in the GIS a web based monitoring system for sugarcane has been prepared. This monitoring system helps the decision makers to take appropriate decisions in order to increase the crop production and other activities related to the crop. The Web GIS helps the nontechnical users to access the information and take appropriate measures to improve the crop production. These user-friendly systems need to be developed and made simple in order to take the technology from the scientific community to the common man.

9.1   INTRODUCTION

Agriculture is the backbone of Indian economy, contributing about 17% towards Gross National Product (GNP) and providing livelihood to about 510% of the population. So, for a primarily agrarian country like India, accurate and timely information on the types of crops grown and their acreages, crop yield and crop growth conditions are essential for strengthening country's food security and distribution system. Pre-harvest estimates of crop production are needed for guiding the decision makers in formulating optimal strategies for planning, distribution, price fixation, procurement, transportation and storage of essential agricultural products. Advance forecasting about crop condition and crop production has a strong bearing on national economy as well as day-to-day life of the masses. The sugar and onion crises of late 90's showed that a faulty crop forecasting system could create havoc for the government and the public.

The space borne remote sensing and GIS technology has been providing information for many agricultural applications, such as identification and area estimation of short duration and marginal crops grown in fragmented land holdings, detection of crop stress due to nutrients and diseases, and quantification of its effect on crop yield. There is a strong need to develop methodologies for improving crop yield models, retrieval of agro-met and canopy parameters, quantification of soil loss, damage assessment and generation of information on moisture and nutrient status of soil subsurface horizons, in order to achieve the best crop yield.

In order to bridge these gaps, the use of WEB GIS and spatial database generated through the high resolution satellite imagery will be used for generating the processing and analytical techniques, that would enable integration of ground based parameters with the satellite data.

In order to study the advantage of using the space technology and information technology in crop monitoring and management of various activities involved in achieving a better yield a study is being done using high resolution satellite data and the use of information technology to create a web GIS application for monitoring the crop production with special emphasis on sugar cane crop.

9.2   SCOPE OF THE STUDY

For the present study a particular land allotted to the sugar plants by the concerned authorities of the government was taken up for micro level planning. The management of the sugar plants is responsible for all the measures for improving the crop production in the allotted land by proving the seeds, fertilizers, pesticides, loans, etc., to the farmers. The management is solely responsible for taking all the measures for improving crop production. In order to achieve the above-mentioned features the management will do precise field wise observation of the land (Fig. 9.1). To monitor each and every field and for better communication from the field data, management will follow the Hierarchy Like Chief Cane Agriculture Officer, Deputy Cane Agriculture Officer, Area Managers, Field Assistants, who will be in-charge for two to three villages. In the absence of an online management system it was observed that gaps were arising in the hierarchy, which May lead to gaps in the communication and management. Due to the wide extent of the area and the hierarchy of employees, it is very difficult to monitor the allotted land very closely. Even for planning the regular water, fertilizers and pesticides applications to the fields and for identifying the optimum way of transportation of the crop which is ready for cutting to the sugar plants is being done manually, which is very tedious and it may lead to improper management, this in turn reduces the crop production. Most of the managements are not keeping the records or history of the fields at one place and all these are operated manually and also not following the technical maps, like soil, slope, and ground water maps to get the more production. Even to monitor the field assistants performance also management don't have any specific system.

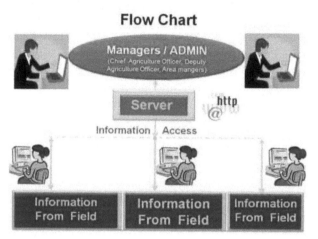

FIGURE 9.1   Flow Chart of Management Information System.

The proposed System will be used for forecasting of crop production, Identifying the fields, which are ready for harvesting, need for the use of fertilizer, and application of water, transportation and canal facilities, etc., for all the fields and also to get field wise reports like delay in providing water to the crops and use of fertilizers in the field at the appropriate time. This technology is effectively used for monitoring the crop production and yield information of crop pertaining to the total land belongs to the client.

The end use of such a system will be felt in obtaining higher productivity, lower overheads, data mining for long-term planning and efficiency improvement, transparency to the management and cost control, enhanced management satisfaction, faster response and optimum utilization of available resources. The remote sensing coupled with GIS techniques play a vital role to achieve the objectives.

9.3   OBJECTIVES

1.   To generate a system to monitor the crop production using Web GIS tools.
2.   To generate the natural resources thematic layers like LULC, ground water, soil and slope for the study area and make use of online study.

9.4 STUDY AREA

A study was carried out in Medak district of Andhra Pradesh. District comprises of 0.96 lakh hectares of forest cover and the net area sown is around 4.0 lakh hectares, in the total cropped area of 5.25 lakh hectares.

9.5 METHODOLOGY

For the present study high resolution Quick bird satellite data of 60 cm resolution is used to prepare the field wise spatial map. The revenue map of the study area is geometrically corrected using the imagery and also by collecting control points using the GPS from the field. The information related to the each survey number is collected by field survey. This data is linked to the spatial data to generate information system. The data generated in the GIS is customized using VB Dotnet and java and a web based spatial decision support system is generated for better planning and management of the harvested crop.

9.6 RESULTS AND DISCUSSIONS

The conventional monitoring system of management can be enhanced with the integration of latest Web GIS technology. The existing conventional monitoring system most of the times relay on the oral communication through telephonic conversations as input for the decision-making. There is large scope for miscommunication of the information or communication gap due to the higher human interference in the system. This miscommunication may lead to delay in decision-making and other works.

The real time information of actual current positions of the fields at various locations is possible with the continuous monitoring with the GIS with supervisors. The proposed system contains two major components of GIS. The static data contains detailed mapping of the allotted field area as a spatial database in WEBGIS platform. The dynamic data regarding the fields of particular sugar plants are collected. The proposed system will take the input and continuously display to enable the management for decision making of the realistic situation. The system can be accessed by separate logins for administrator and user. There is a scope for unlimited number of users to login to the system. By logging in the admin we can have a total view of the area allotted for the particular sugar plant. The administrator has the rights to view a supreme authority, from this view and creating new ID for managers, filed supervisors and other authorized persons, etc.

9.7   PROPOSED APPROACH FOR SYSTEM

Generally in every sugar industry the typical organization hierarchy followed is given in the following subsections.

9.7.1   ADMIN LOGIN

From the Given WEB Site ADMIN can Login like,
1.   New User ID Creation: Any number of new user ids can be created by using administrator log in.
2.   Field Supervisor Login

The total spatial view of the area allotted for the Supervisor of the Sugar Plant can be seen after logging in to the system. From the spatial view the information about each individual field like name of the farmer, date of sowing, date of application of fertilizer and water, other crop information, etc., which is useful for system to avoid any dependency on the oral communication. Thus, the human error in communication can be minimized with the continuous monitoring of real time data available within the system by any manager prior to taking major decisions.

9.7.2   FIELD HISTORY

From the SDSS the history related to every field can be analyzed. For example, the date on which water and fertilizer need to be applied can be known for each field. If there is a delay or laxity by the farmer in the applying of water or fertilizer the same can be known from the SDSS package. It also provides information about the date on which the field needs to be watered and also the due date for which the field should be watered. In case of the application of the fertilizers to the field can also be known using the SDSS package.

9.7.3   MANAGER LOGIN

The total view of the area allotted for the sugar plant can be seen after login in the system. The road network of the study area is captured from the satellite imagery and is overlaid on the map. From this we can do analysis on the nearest route to reach the particular field and the SDSS can also predict alternative route in case of any problem in the main route.

Similarly from the SDSS package we can identify the fields which are ready for cutting and to view the delay reports of water, fertilizers, etc., from the fields and then see and rectify the delays by avoiding the dependency in the oral communication by seeing the data about each and every field like Farmer name, date of seeding, date of application of fertilizer and water, other crop information and, etc., which is useful for continuous monitoring of fields to know the status of the crop and also the performance of the field supervisor. Thus, the human error in communication can be minimized with the continuous monitoring of real time data available with in the system by any superior prior to implement the decision.

9.7.4 DELAY REPORTS

1. Delay in Water Application
2. Delay in Fertilizer Application
3. Field ready for Cutting

9.8 FACILITIES PROVIDED BY DECISION SUPPORT SYSTEM FOR AGRICULTURE CROP MONITORING

- Easy to use and access the data.
- Getting online information about fields and efficiency of the fields yielding, etc.
- Easy to get the performance of filed supervisors.
- Easy to get the history of each and every field.
- Giving feel like actual field/site visit to the management.
- Reduce the delay in application of water, fertilizers, etc.
- Transparency in crop growth and land activities with higher productivity and lower overheads.
- Transparency to the management and cost control.
- Transparency for planning of application of water from canals, route planning for transportation of crop after cutting to the sugar plant, etc.
- Easy to maintain the fields belong to particular sugar plants.
- Easy to get the status of each and every field, so that total allotted land status of each sugar plant and every stage of the crop, route map to reach the location.
- Faster response and optimum utilization of available resources.
- Getting the online status of every stage of the crop in particular field.

9.9 SUMMARY OF DELAY REPORT TO SUGAR PLANT MANAGEMENT

The system provides continuous monitoring on crop activities and facilitates dynamic planning of strategies to meet the changing requirements of the crop or sugar plant activities with respect to the impedance of the implementation of planned strategies. The scope for the enhancement in the system is much more as the advancement of the web technology. The databases can be made available for all the concerned authorities through Internet and Intranet facilities. This connectivity definitely enhances the mutual cooperation of the authorities to accomplish the task with more efficiency.

9.10   NEED FOR THE SYSTEM

The proposed System will be used for:
1. better operation and maintenance of sugar plant management.
2. acts as a better decision tool for management of crops.
3. to maintain the records available at different hierarchy at one place with better record keeping.
4. enhance the decision making better and transparent.
5. for online maintenance of sugar plant.
6. to access and use of the data easily.
7. to get online information about fields and efficiency of the field operations, etc.
8. to get the performance of filed supervisors.
9. to get the history of each and every field.
10. to give feel like actual field/site visit to the management.
11. to reduce the delay in application of water, fertilizers and, etc.
12. to get the transparency in progress of crop growth.
13. to get higher productivity and lower overheads.
14. to get the transparency in total land activities.
15. to give transparency for planning of application of water from cannels, route planning for transportation of crop after yielding to the sugar plant and, etc.
16. to get the status of each and every filed, so that total allotted land status of each and every stage of the crop.
17. to get the route map to reach the location.
18. to get faster response and optimum utilization of available resources.
19. to get the online status of each and every stage of the crop in particular field.
20. to get the delay reports from filed about field activities.

9.11   CONCLUSION

WEBGIS is essential to effective preparedness, communication, and training tool for sugar plant management, which offers enhancement for the existing manual system to avoid the human error in making the decisions to enhance the crop production and also provides continuous monitoring of the crop and field activities and operations. With the effective utilization of the advantages provided by the WEBGIS spatial database the efficiency in the planning and implementing activities can be improved.

KEYWORDS

- **Agriculture**
- **Agriculture Crop Monitoring System**
- **Geographic Information System**
- **GIS Technology**
- **Gross National Product**
- **Quick Bird Satellite Data**
- **SDSS Package**
- **Sugar Plant Management**
- **Sugar Plant Management**
- **Sugarcane**
- **Web Based Monitoring System**
- **Web GIS**

REFERENCES

Anselin, L. (1990). What is special about spatial data Alternative perspective on spatial data analysis. In Spatial Statistics: Past, Present and Future, edited by D. A. Griffith, Monograph #12, Institute of Mathematical Geography.

Fundamentals of Geographic Information Systems- Lillisand and Keefer.

GIS Development magazines of August 2003 and October 2003. Upton, G. J. Fingleton, B. (1985). Spatial data analysis by example, volume 1: Point pattern and quantitative data. Wiley, Toronto Singapore, Brisbane.

CHAPTER 10

SITE SUITABILITY ANALYSIS FOR WATER HARVESTING STRUCTURES IN WATERSHED USING GEO-INFORMATICS

N. S. R. PRASAD[1], P. KESEVA RAO[2], R. R. HERMON[3], and V. MADHAVARAO[4]

CONTENTS

[1] Assistant Professor, CGARD, NIRD & PR
[2] Assistant Professor CGARD, NIRD & PR
[3] Associate Professor CGARD, NIRD & PR
[4] Professor and Head CGARD, NIRD & PR

ABSTRACT

Since India is one of the developing countries, due to increased urbanization and deforestation the vegetative cover is being decreased as a result of which the hydrological cycle is being affected and this is resulting in bottoming levels of the ground water. All these consequences finally result in desertification of that piece of land. Watershed planning is very much essential for the conservation of land and water resources and their management for optimum productivity.

The efficient and proper utilization of water resources is very essential to fulfill the water requirements for various purposes like drinking, irrigation and industrial use and various other household activities. This can be achieved by proper planning of a watershed. Conventional and manual techniques for the study of watershed characters are very expensive and time consuming. Remote sensing and GIS techniques can provide quick and accurate information about watershed characteristics for efficient planning of a particular watershed.

10.1 INTRODUCTION

Water resources are increasingly in demand in order to help agricultural and industrial development, to create incomes and wealth in rural areas, to reduce poverty among rural people, and to contribute to the sustainability of natural resources and the environment. Reliable and timely information on the available natural resources is very much essential to formulate a comprehensive land use plan for sustainable development. The land, water, minerals and biomass resources are currently under tremendous pressure in the context of highly competing and often conflicting demands of an ever-expanding population. Consequently over exploitation and mismanagement of resources are exerting detrimental impact on environment.

In India more than 75% of population depends on agriculture for their livelihood. Agriculture plays a vital role in our country's economy. In order to mitigate droughts which occur frequently in several parts of the country especially in dry land areas the Ministry of Agriculture and Cooperation has launched an integrated watershed concept using easy, simple and affordable local technologies. Watershed approach has been the single most important landmark in the direction of bringing in visible benefits in rural areas and attracting people's participation in watershed programs (Singh et al., 2005). The programs under watershed approach broadly fall into soil and water conservation, dry land and rain fed farming, ravine reclamation, control of shifting cultivation and improvement in the vegetative cover. The basic objective is to increase production and availability of food, fodder and fuel; restore ecological balance. Development of the watershed needs better understanding about the various natural resources, their relations with each other and their relations with livelihood of the stakeholders.

Remote Sensing (RS) data and Geographical Information System (GIS) play a rapidly increasing role in the field of land and water resources development. One of the greatest advantages of using Remote Sensing data for natural resource management is its ability to generate information in spatial and temporal domain, which is very crucial for successful model analysis, prediction and validation.

The present study is an attempt using Remote Sensing and GIS techniques to propose various water harvesting and soil conservation measures in order to suggest water resource development plan for Sajjaram Tanda watershed covering 800 ha in Zahirabad mandal of Medak district in Andhra Pradesh.

10.2   LOCATION AND EXTENT

The location of study area Sajjaram Tanda is situated at a distance of 12 kms from Zahirabad Town (Fig. 10.1). The study area falls in Survey of India Topo sheet No.56G/10SW (Fig. 10.2). The area experiences semiarid type of climate.

FIGURE 10.1   Map of Study Area.

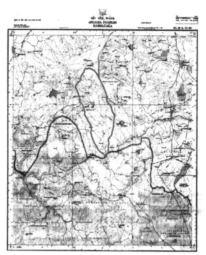

FIGURE 10.2   Extent of Study Area.

10.3   SCOPE OF THE STUDY

In large number of cases the failure of watershed development is largely due to mal-adjustments with diverse facets of nature, caused by lack of awareness of the natural resources. In order to generate optimum utilization of existing natural resources like land, vegetation and water in watershed proper scientific surveys should be conducted. Geo-scientific studies of the terrain, socioeconomic appraisal of the stake holders and the use of Remote Sensing Data for faster assessment of natural resources such as soil, geology, drainage, etc., as well as assessment of economic activities through land use and infrastructure of the watershed area is well known. This is also used for monitoring of watershed development at later years. GIS is a very powerful tool for development of the watershed area with all natural and socioeconomic facets for better planning, execution and monitoring of the project.

10.4   OBJECTIVES

The main objective of this chapter is to generate information/databases on 1:25,000 scale pertaining to hydro-geomorphology, Drainage, surface water bodies, watershed, transport network, etc., using multitemporal satellite data. Conversion of these databases into digital form for future analysis and utilization and to prepare water resources development plan, such that control of soil & moisture conservation and land degradation, conservation and management of water resources can be achieved.

10.5   METHODOLOGY

To achieve the above objectives, the following methodology (Fig. 10.3) and proce-dure is adopted in the present study. Collection of satellite data and Survey of India Topographical maps covering the study area, prepared of Base map (Fig. 10.4) on 1:25,000 scale using Survey of India Topographical Maps, prepared of Drainage (Fig. 10.4), watershed and Surface water bodies using SOI topographical maps, pre-pared (Fig. 10.4) of 10 m contour interval using SOI topographical maps, prepared Triangulated Irregular Network (TIN) map (Fig. 10.4) from contours and then pre-pared of Digital Elevation Model (DEM) (Fig. 10.5) from TIN, prepared of Slope (Fig. 10.5), Aspect (Fig. 10.5), Flow Accumulation (Fig. 10.5) maps from DEM, prepared Geomorphology (Fig. 10.6), Ground water prospecting map (Fig. 10.6), Land use/Land Cover (Fig. 10.6) using Satellite data on 1:25,000 Scale after ground truth data collection, verification of doubtful areas and correction, modification and transfer of post field details and recommendation of water resources development plan.

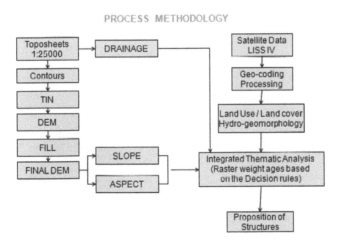

FIGURE 10.3   Process Methodology.

In this chapter, most of the analysis was done using hydrological analysis tools in ArcGIS 9.2 software. The hydrological analysis process in GIS is one of the ef-fective methods in terms of cost and time in proposing various water harvesting structures. This process deals with assessing various hydrological characteristics of a surface. Slope and aspect play a vital role in determining the shape of a surface. The basic inputs required to generate a hydrological model for a region are slope, aspect, flow direction, flow accumulation, and a possible stream network.

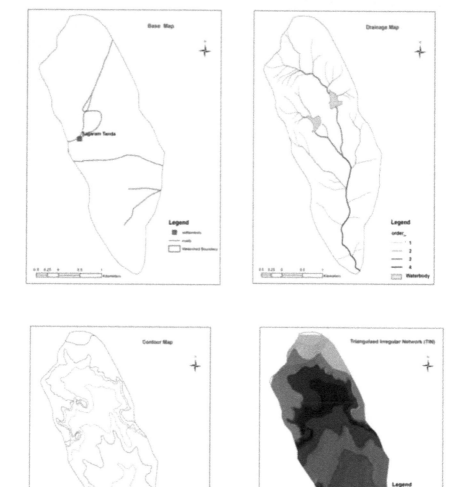

FIGURE 10.4   Maps of Base, Drainage, Contour and TIN.

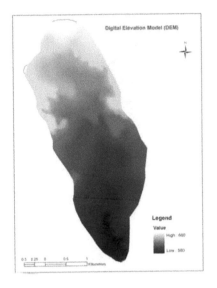

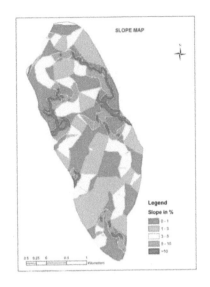

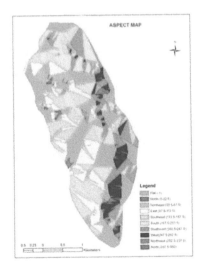

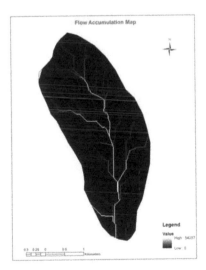

FIGURE 10.5   Maps of DEM, Slope, Aspect and Flow Accumulation.

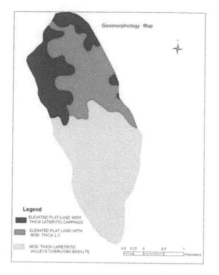
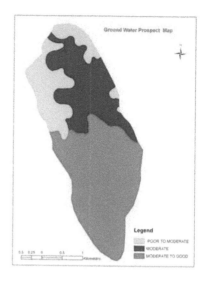

FIGURE 10.6   Maps of Geomorphology, Ground Water prospects, Land Use and Land Cover and Land and Water Resource Development Plan.

Water resource development plan has been prepared on the basis of integration of information on geomorphology and land use/land cover, drainage, ground water prospect map and slope. Suitable structures are suggested for surface harvesting/ recharge. Weightages are given to significant units (on priority basis) in various thematic layers such as ground water prospects, slope, drainage and land use/land cover in raster form in order to prioritize locations for suggesting appropriate recharge structures. First, all the thematic layers have been converted to raster form to assign weightages, since the analysis should be performed in raster mode. For suggesting check dams weightages given in ground water prospecting, high priority is given to Moderate to good, medium priority is given to moderate g/w prospects and least priority is given to poor to moderate g/w prospects. In the land use/land cover layer high priority is given to Fallow areas, moderate priority to croplands and least priority to scrub lands. Similarly, the slope classes have been reclassified into 5 categories according to AISLUS classification. Slope 1–3% is given high priority; slope less than 1% and more than 3% has given less priority. In stream order raster 2nd and 3rd order streams are given high priority, 1st and 4th order streams are given least priority. Final Weightages were given among the thematic layers using raster calculator in spatial analyst of ArcGIS, based on the Weightages priority raster for check dam is generated. After verifying in the field, the structures were finalized. Same procedure applied for the different water harvesting structures, based on the priorities.

Based on this location priority, various water-harvesting structures have been proposed (Fig. 10.6). All these functions are used either by themselves or in conjunction with other mathematical or statistical or simulation models for varied applications in GIS. There is a wide range of spatial analysis functions in a GIS. To take advantage of its capabilities, user must define the problem clearly; decide the data required and the spatial operations to be performed for reaching the goal.

10.6 CONCLUSION

Generated information/databases on 1:25,000 scale pertaining to base map, drainage, contours, hydro-geomorphology, ground water prospects, land use/land cover, slope and aspect using multitemporal satellite data and SOI toposheets.

Proposed various water harvesting structures like check dams, contour Bunding, continuous contour trenches and vegetative barriers at appropriate places by assigning the weightages based on the priorities using hydrology tools in ArcGIS software. This model is very useful for the site suitability analysis for water harvesting structures in watershed development projects.

This hydrology modeling enables us to better planning in watershed development by giving the scientific inputs. This enables to create the digital database for future use and save the time and investment for preparation DPR/action plan.

KEYWORDS

- **Digital Elevation Model**
- **Geographical Information System**
- **GIS Technique**
- **Hydrological Analysis Process**
- **Hydrological Analysis Tools**
- **Remote Sensing**
- **Triangulated Irregular Network**
- **Watershed Planning**

REFERENCES

Ft. Lauderdale. Sander, P. (1997). Water-Well Siting in Hard- Rock Areas: Identifying Promising Targets Using a Probabilistic Approach, Hydrogeology Journal, 5(3), 32–43.732.

Gustafsson, P. (1993). High-resolution satellite imagery and GIS as a dynamic tool in groundwater exploration in a semiarid area.

Krishnamurthy, J., Kumar Venkates, N., Jayaraman, V., Manivel, M. (1996). An approach to demarcate groundwater potential zones through remote sensing and a geographic information system, International Journal of Remote Sensing, 17(10), 1867–1884.

Minor, T., Carter, J., Charley, M., Knowks, B., Gustafson, P. (1994). The use of GIS and Remote Sensing in Groundwater exploration for developing countries. Proc. of the 10th ERIM thematic conference on geologic remote sensing held in San Antonio, USA on may 9–12 1994 pp.168–179.

Proceedings of AWRA symposium on GIS and Water Resources held in Ft.Lauderdale, Florida, USA, 1996. Florida.

Proceedings of the IAHS conference on Application of Geographic Information Systems in Hydrology and Water Resources Management held in Vienna, Austria, on April 19–22 1993, 93–100. Vienna.

Richards, C. J., Roaza, H. P., Pratt, T. R. (1996). Applying geographic information systems to groundwater assessments.

CHAPTER 11

GENERATION OF SUSTAINABLE LAND USE PLANS USING GEO-INFORMATICS TECHNOLOGY

PHANINDRA KUMAR[1], V. MADHAV RAO[2], P. KESAVA RAO[3], and R. R. HERMON[4]

CONTENTS

[1]Sr. Consultant, CGARD, NIRD & PR
[2]Professor & Head CGARD, NIRD & PR
[3]Assistant Professor CGARD, NIRD & PR
[4]Associate Professor CGARD, NIRD & PR

ABSTRACT

For optimal utilization of available natural resources and for taking up any preventive or curative measures, timely and reliable information on natural resources with respect to their nature, extent and spatial distribution; and nature, magnitude and temporal behavior of various types of degraded lands, is a prerequisite. Hitherto, such information has been generated through conventional approach using topographical sheets or aerial photographs. Remote sensing has provided opportunity with Synoptic view of a fairly large area with multispectral measurements at regular intervals and enabled generating information on natural resources, in a timely and cost-effective manner. The objective of the present study is to generate information/ databases on 1:50,000 scale pertaining to drainage, surface water bodies, watershed, transport network, land use land cover, hydrogeomorphology, slope, soil and its related parameters like soil depth, water depth, gravelliness, etc., using multitemporal satellite data. A scheme for thematic data integration and recommendation for various combinations of land parameters was evolved by observations in the field. Following the scheme of data integration, action plan maps were generated giving suitable site – specific recommendations for alternate land use and water conservation measures. The Land Resource Information System (LRIS) was designed for all the thematic layers generated in order to enable the users to work without help of traditional costly software's.

11.1   INTRODUCTION

Till the recent past, land was looked in a narrow perspective of being a physical entity in terms of its topography and spatial nature. But the broader, integrative or holistic view, takes into account both, a vertical aspect – from atmospheric climate down to ground water resources, and a horizontal aspect an identifiable repetitive sequence of soil, terrain, hydrological, vegetative and land use elements. Soil, water, flora and fauna are the important land resources, which together influence the survival of human beings by supporting food production and providing a congenial living environment. There was harmony between the mankind, the other living beings and the surrounding environment. Increasing needs of the growing population inevitably lead to agriculture and industrial development through haphazard usage of land resources, which eventually disturbed the ecological balance through pollution and declining quality of soil, water and atmosphere. In developing countries, and especially in the most densely populated countries like India, there is an urgent need for immediate action as there will be pressure to make ad-hoc decisions and to push forward development schemes, for social or political reasons.

For rational utilization of available natural resources and for taking up scientific decisions timely and reliable information on natural resources with respect their nature, extent and spatial distribution; and nature, magnitude and temporal behavior of various land use features is required. Hitherto, such information has been generated through conventional approach using topographical sheets or aerial photographs. Remote sensing has provided opportunity with Synoptic view of a fairly large area with multispectral measurements at regular intervals and enabled generating information on natural resources, in a timely and cost-effective manner.

11.2 LAND RESOURCES INFORMATION SYSTEM

Advances in geo-information technology have created exciting possibilities of collecting, and managing large amounts of data from earth resource processes in various forms and scales. Also social and economic data is increasingly available through census, development and health surveys, etc. For improved decision-making, the required information, tools, techniques, models and decision-making procedures have to become integrated in a user-friendly information processing system called "Land Resources Information System." In spatial modeling, multiple criteria methods allow for the presence of more than one objective or goal in a complex spatial problem. However, they assume that the problem is sufficiently precise that the goals and objectives can be defined exactly. Such problems require a flexible approach. The system should assist the user by providing a problem-solving environment. LRIS generally provides a framework for integrating:

1. analytical modeling capabilities;
2. database management systems;
3. graphical display capabilities;
4. tabular reporting capabilities;
5. the decision-makers expert knowledge.

11.3 LOCATION AND EXTENT

Zaheerabad mandal lies in the Medak District, which is situated in the Telangana region of Andhra Pradesh, India. Zaheerabad is located about 100 km from Hyderabad on the Mumbai-Hyderabad national highway (NH-9). The mandal is bounded on the north by Nyalkal, North-east by Jharasangam, South-cast by Kohir mandals of Andhra Pradesh and Gulbarga and Bidar of Karnataka state in west. The mandal lies between North Latitudes 17° 46' 22" and 17° 32' 32" and East Longitudes 77° 26' 42" and 77° 42' 36" and falls in Survey of India Toposheet Nos. 56G/6, 56G/9 and 56G/10 of scale 1:50,000 published by Survey of India in 1973. The total area of the study area is 392.15 sq. km.

11.4 OBJECTIVES

The main objective of the present study is to generate information/databases on 1:50,000 scale pertaining to drainage, surface water bodies, watershed, transport network, land use land cover, hydrogeomorphology, slope, soil and its related parameters like soil depth, water depth, gravelliness, etc., using multitemporal satellite data. These databases are converted in to digital form for future analysis and utilization and to prepare location specific land and water resources development plans, by integrating with socioeconomic data and contemporary technology in the GIS environment such that control of soil and moisture conservation and land degradation, optimal management of croplands and conservation and management of land and water resources can be achieved.

11.5   METHODOLOGY

The following methodology is adopted in the present study to meet the above-mentioned objectives. The base map is generated at 1:50,000 scale from the SOI Toposheet. The thematic layers like LULC, hydrogeomorphology, soil, slope, etc., are generated using the IRS P6 LISS IV images. Taking the SOI Toposheets as source, the thematic layers like drainage and contours are prepared at 1:50,000 scales. The slope map is derived using Survey of India topographical sheets at 1:50,000 scale with 20-meter contour interval. The rainfall and temperature data and other collateral data of the study area are collected and integrated in the GIS Domain. The digital elevation model (DEM) is generated from the contours. The soil map is taken as the base for integration.

A scheme for thematic data integration and recommendation for various combinations of land parameters was evolved by observations in the field. Following the scheme of data integration, action plan maps were generated giving suitable site – specific recommendations for alternate land use and water conservation measures. While formulating the locale-specific action plan, the earlier research carried out by various research centers in the field of agriculture, etc., were taken into consideration along with the prevailing socioeconomic conditions. All these data, which are generated in the GIS domain are used as inputs in the LRIS developed using the map objects and VB. The LRIS is an package which works independent of the GIS software's and tools were created in the package in such a that any use user without the technical knowledge can also operate the package and can perform some of the functions like displaying the data, querying the data, overlay and union which are there in the GIS software.

11.6   RESULTS AND DISCUSSIONS

Sustainable agricultural land use and cropping pattern plans of the area are generated by GIS based logical integration of crop suitability's, land productivity, land capability, socioeconomic and terrain characteristics information. Specific action plans are devised for optimum management of land and water resources through integration of information on natural resources, socioeconomic and meteorological data and contemporary technology. An important task of LRIS is to facilitate the linkage between nonspatial data and the spatial data. The nonspatial data like census, socioeconomics, agricultural inputs, etc., would be available in tabular form and has to be retrieved from the relational database management system (RDBMS). On the other hand, the spatial information is in the form of maps, referenced to the geographic latitudes-longitudes. The nonspatial tabular information is linked to the spatial information through a customized GIS approach. This query-shell facilitates data handling. The strength of GIS is the integration of multilayered data from different sources and various scales. The integration of different layers of information has been a difficult task manually until the maps were drawn on a transparent film. With the availability of GIS, which takes the data into digital space, the ability to see through maps, which are overlaid one over the other digitally and analyze the maps is achieved. Database management systems integrated with graphic interface have a powerful query capability. This will finally give the analytical ability to pose complex query and extract information spatially.

11.6.1   LAND USE LAND COVER

The knowledge of spatial distribution of land cover/land use of large area is of great importance to regional planners and administrators. Conventional ground methods are time consuming and no uniform classification system was used in the preparation of maps with the advent of remote sensing technology the above problems have been solved to quite some extent. Satellite data can provide information on large areas and the temporal data can be used for change detection and updating old data. The land use/land cover categories that can be obtained from the remotely sensed data include level I classes of land use classification system such as water bodies, forest, grass land, agricultural land, barren land, and scrub land. The Spatial Distribution of the various land use land cover classes found in the study are given below in the Table 11.1.

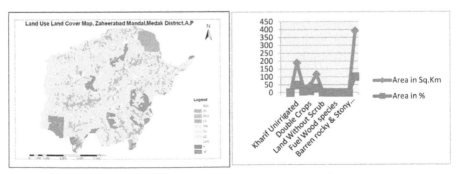

FIGURE 11.1   Land Use Land Cover Map of Zaheerabad Mandal.

TABLE 11.1   Area Extent of Land Use Land Cover Classes

Land use/Land cover category (LULC)	Area in Sq.Km	Area in %
Forest Plantation	3.00	0.76
Kharif Unirrigated	188.83	48.13
Scrub Forest	31.20	8.00
Kharif Irrigated	32.12	8.28
Rabi Irrigated	10.24	2.62
Land with Scrub	115.14	29.3
Land Without Scrub	5.71	1.5
Deciduous Forest	3.52	0.9
Fuel Wood species	0.96	0.2
Plantation	1.27	0.3
Barren rocky and Stony waste	0.04	0.01
Total	392.13	100

11.6.2   GEOLOGY

The mandal is characterized by gently undulating topography with low hills, the highest elevation being 660 m above msl at the central parts of Zaheerabad and lowest 540 m above msl at the southern part of this mandal. Highest elevation is extending from Gopanpalli in the central to Asadganj in the north; other one extends from Mugdampalli in the south east to the Dhanaseri in the west in the district. Pre-cambrian rocks, such as granite, adamellite, tonalite, amphibolite, homlende boitite schist occupy a major part of the mandal. These formations were subjected to tectonism and greenschist facies metamorphism. Except for a portion in the western part of the mandal, most of the area is occupied by granites. All these rock formations are traversed by NE-SW and N-S trending dolerite dykes and vien quartz. The Deccan Traps and intertrappeans are seen at the central part of the mandal, extending up to the northern and western border of the mandal. The laterite copings are seen extensively around Zaheerabad. Two major lineaments trending NE-SW direction from Hotibuzurg to Haiderabad and another one trending in the same direction near to Zaheerabad occur.

11.6.3   GEOMORPHOLOGY AND GEOHYDROLOGY

The mandal is covered under Godavari basin. Karanja Vagu, is a major small river which flows NW – NE direction in this mandal. The drainage pattern is dendritic in granite terrain and parallel in Deccan Traps. The hard rocks viz., Archeans and Deccan Traps occupy 79.35%, soft rocks viz., laterites 20.63% and alluvium 0.2%.

The depth to water table in granite varies from 4.5 to 199 mm, in laterites from 5 to 30 m and in alluvium from 0.5 to 7.5 m. In Deccan Traps, Water occurs in fractures, crevices and joints. A number of lineaments form the promising zones for ground water development. Laterites are highly porous and permeable and possess good water bearing capacity. Ground water is moderately hard and is required to be softened before use. The land forms in the district are mostly structural, denudation and fluvial in origin. In the northern parts, crystalline complex represented by meta sediments, gneisses and granite forms a distinct pediplain. This unit resulting out of the denudational process comprises landforms like residual hills, Inselberg and shallow weathered pediplain. In the western part, landforms are of moderately dissected plateau.

11.6.3.1 HYDROGEOMORPHOLOGY

Hydrogeomorphology deals with the study of landform in relation to groundwater occurrence and availability. It is manifested at the surface, mainly by geology, geomorphology, structure and recharge conditions. All the four parameters were studied and integrated to arrive at the groundwater prospects under each geomorphic cum lithological unit, designated as hydrogeomorphic unit. The following Geomorphic units are mapped in the Zaheerabad Mandal at 1:50,000 scale and presented in Table 11.2.

TABLE 11.2 The Various Geomorphic Units, which are Present in Zaheerabad Mandal

S. No.	Map Unit	Geomorphic Unit	Litho Unit	Structure	Description
1	FV	Filled-in-Valley (FV)	Latcrrite	Concretionery detrital laterite resting over the lithomargic clays	Broad valleys formed due to removal of hard crust of laterite consisting of detrital/concretionary laterite.
2	MDP	Highly Dissected Plateau (MDP)	Deccan Trap (Volcanic Flows)	Flat to gently dipping volcanic flows (10–20 m thick each). Individual flows are hard and massive at bottom grading to vesicular or tuffaceous at top. Thin clay/grey earth beds mark the contacts between the flows.	Upland areas dissected by 5–15 m deep narrow valleys.

3	P1	Pediment Insel-berg Complex (PI)	Penin-sular Gneiss and Granite	Massive granite and foliated gneiss, traversed by joints, fractures and faults.	Gently undulating rock cut plain with small hills and rock out crops.

11.6.4   GROUND WATER POTENTIAL MAP

Ground water potential maps are prepared by integrating information on geomorphology, slope lithology, structural features and the precipitation. Ground water recharge depends on favorable slope, permeability and degree of compactness of the rocks. The movement and occurrence of ground water is controlled mainly by geology, geomorphology and recharge conditions of the area. The geology and geomorphology of the study area have been studied and by combining the individual Litholo-landform units the geomorphology map is prepared. These geomorphic units have been evaluated for their ground water prospects based on the hydrogeological characteristics of the geological and geomorphological parameters. The various ground water potential zones present in Table 11.3.

Brief description of various ground water potential zones in Zaheerabad Mandal: salient features of various Ground water potential zone units, delineated in the Zaheerabad Mandal are given hereunder.

Zone I: It consists of geomorphologic units like Filled in Valley. The Ground Water Potential is good in this Zone. The material in Filled in valley 1 is made up of flat to gently dipping volcanic flows of 10–20 cm thickness each. The individual flows are hard and massive at bottom grading to vesicular at the top. Thin clay earth beds mark the contact between the flows. These are gently undulating wide valley floors with good recharge from MDP2, HDP2, and SDP2. The ground water potential in the weathered zones and also along faults/fractures suitable for dug wells and bore wells, respectively. The contact zones between trap and granitic basement often form good aquifers. The filled in valley is made up of concretionary detrital literate resting over the lithomargic clays. These are broad valleys formed due to removal of hard crust of laterite consisting of detrital laterite. It is suitable for good yields suitable for dug, bore wells, tube wells. The potential will be better along faults and fractures. (Note: Zone II is negligible area and therefore it is not considered in this chapter).

Zone III: It consists of geomorphologic units like Moderately Dissected plateau. The ground water potential in this zone is Moderate.

Moderately Dissected Plateau (MDP2): The lithographic unit is Deccan Trap (Volcanic Flow). These are Upland areas, which are dissected by greater than 5-meter deep narrow valleys. These are flat to gently dipping volcanic flows with a depth of 10–20 meters thickness each. Individual flows are hard and massive at bottom grading to vesicular at top. The clay earth bed marks the contacts between the flows.

It has limited to moderate potential in the weathered zones along narrow valleys. It has moderate to good potential suitable for deep bore wells along the fault and fracture zones.

Zone IV: It consists of geomorphologic units Pediment Inselberg Complex. The ground water potential in this zone is Moderate to Poor.

Pediment Inselberg Complex (PL4): The lithographic unit is peninsular gneiss and granite. These are gently undulating rock cut plains with small hills and rock outcrops. These are made up of massive granite and foliated gneiss, which are traversed by joints, fractures and faults. They have moderate potential along faults and fractures, which are suitable for borewells subject to recharge.

The areal extent of the various geomorphic units with the ground water potential for the zaheerabad mandal is given in Table 11.3.

TABLE 11.3 Area Extent of Geomorphic Units of Zaheerabad Mandal

Geomorphic Units	Ground water Potential	Area in Sq.Km	Area in %
Moderately Thick Lateritic Plateau	Moderate	84.05	21.43
Moderately Thick Lateritic Valley	Moderate to Good	141.80	36.17
Thick Lateritic Plateau	Moderate to Poor	166.28	42.40
Total		392.13	100.00

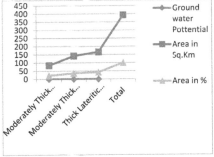

FIGURE 11.2 Hydro Geomorphology Map and Various Ground Water Potential Zones of Zaheerabad Mandal.

11.6.5 SOILS

The district exhibits a variety of soils (Fig. 11.3). Black cotton soil ranging in thickness from 0.3 to 0.75 m occurs in Zaheerabad, Tumkund, Kotur and Didgi villages. Red soil ranges in thickness from 0.75 to 2 m. It possesses high water retention capacity. Major part of the district is arable unirrigated land around Zaheerabad. Around 17 soil series were mapped for the entire mandal using the satellite imageries and field survey at 1:50,000.

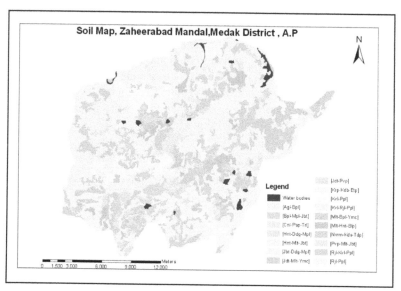

FIGURE 11.3   Soil Map of Zaheerabad Mandal.

11.6.6   DERIVED SOIL PARAMETERS

Soil-site and other land characteristics studied and recorded during field data collection and mapping have been interpreted and the soil map units grouped with respect to the major soil attributes such as soil depth (effective rooting depth), surface soil texture, gravelliness and stoniness, soil drainage, soil available-water capacity, soil slope, soil erosion, calcareousness, etc. Such interpretative grouping of soil and land characteristics help in identifying areas that have specific problems like shallow rooting depth, severe erosion, poorly drained areas, sandy areas, steeply sloping lands, salt affected lands, etc., and areas that have high potential like very deep soils, nongravelly soils, nonsaline or level areas, etc., for sustained agricultural and nonagricultural uses.

11.6.6.1   SOIL DEPTH

Soil depth determines to a great extent the rooting system of plants which is ultimately reflected in crop growth and crop yield. It determines the capacity of the soil column to hold water and supply plant nutrients in relation to soil texture, mineralogy and gravel content. The extent of area under each depth-class association for the study area is given in Table 11.4 and Fig. 11.4.

TABLE 11.4   Area Extent of Soil Depth Units of Zaheerabad Mandal

Mapping Unit	Description	Area in Sq. Km	Area in %
2	Extremely Shallow	98.99	25.24
3	Shallow	79.52	20.28
4	Moderately Shallow	17.36	4.43
7	Very Deep	192.10	48.99
	Water Bodies	4.16	1.06
	Total	392.13	

 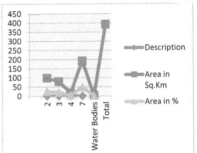

FIGURE 11.4   Area Extent of Soil Depth Map of Zaheerabad Mandal.

11.6.6.2   SOIL EROSION

Soil erosion refers to the wearing away of the earth's surface by the forces of water, wind and ice. It is both destructive and constructive. It is responsible for causing variations in topography by eroding elevated land surfaces and at the same time depositing the eroded material in the plains, basins and valleys. It is further aggravated by human intervention through indiscriminate felling of trees, overgrazing, mining, and cultivation on steep slopes and degraded lands, thus affecting the natural ecosystem. Based on the intensity of erosion as observed visually during field survey and mapping, and also the amount of soil (loss of A or B horizons) removed in the soil profiles examined, the soils of study area have been classified under three erosion classes 'liz.;- ej-nil or slight erosion, e2-moderate erosion and e3-severe erosion.

11.6.6.3   SOIL CALCAREOUSNESS

Calcareousness (lime content) is the term used to indicate the content of calcium carbonate in the soil. In the field, it is estimated by observation of the effervescence given by the soil when it is moistened with dilute hydrochloric acid. Soil calcareousness classes used were 0 – nil effervescence (noncalcareous), 1 – slight effervescence (slightly calcareous), 2 – strong effervescence (strongly calcareous). Calcareousness influences soil pH and the availability of macronutrients and micronutrients. Physical conditions of soils are also greatly influenced by the quantity and size of lime nodules and concretions.

The extent of area under each calcareousness class association for the study area is given in Table 11.5.

TABLE 11.5   Area Extent of Soil Calcareousness Units of Zaheerabad Mandal

Mapping Unit	Description	Area in Sq. Km	Area in %
0	0–15% Non Gravelly	76.04	19.39
1	15–35% Slightly Gravelly	117.365	29.93
2	35–60% Gravelly	194.58	49.62
	Water Bodies	4.16	1.06
	Total	392.13	100.00

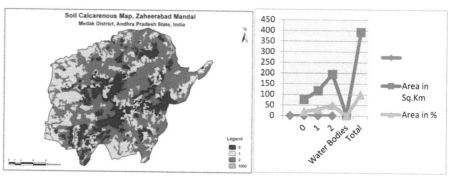

FIGURE 11.5   Area Extent of Soil Calcareousness Units of Zaheerabad Mandal.

11.6.6.4   SOIL AVAILABLE-WATER CAPACITY

Available Water Capacity (AWC) of soils is mainly dependent on the amount, intensity and distribution of rainfall, infiltration, permeability, drainage and texture, type of clay minerals, soil depth and content of coarse fragments. Classes of soil available-water capacity are based on the ability of soil column to retain water between the tensions of 0.33 kPa and 15 kPa in a depth of 100 cm soil or the entire column if the soil is shallower. The AWC of soils of the study area estimated from soil depth, gravel and stone and mineralogy can help in determining the length of crop growing period which helps in land use planning. The AWC classes used for grouping the soils of the district were (1) very low (<50 mm m^{-1}), (2) low (50–100 mm m^{-1}), (3) medium (100–150 mm m^{-1}), (4) high (150–200 mm m^{-1}), and (5) very high (>200 mm m^{-1}). The extent of area under each AWC class association for the study area is given in Table 11.6.

TABLE 11.6   Area Extent of Soil Water Capacity Units of Zaheerabad Mandal

Mapping Unit	Description	Area in Sq.Km	Area in %
1	Very Low	189.17	48.24
2	Low	44.28	11.29
5	Very High	154.52	39.41
	Water Bodies	4.16	1.06
	Total	392.13	100.00

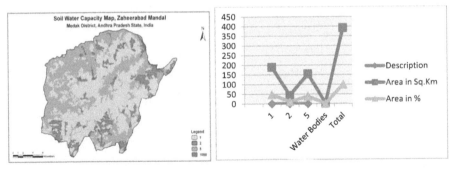

FIGURE 11.6   Area Extent of Soil Water Capacity Units of Zaheerabad Mandal.

11.6.6.5   GRAVELINESS AND STONINESS

Gravel is the term used for describing coarse fragments between 2 mm and 7.5 cm diameter and stones between 7.5 cm to 25 cm. The presence of gravel and stones in the soil reduces the volume of soil that affects moisture storage, drainage, infiltration and runoff, and hinders plant growth by impeding root growth and seedling emergence, intercultural operations and farm mechanization. The gravel and stone content by volume for each of the soil horizon, as well as on the surface recorded during soil survey were used for grouping the soils into different gravelly or stony classes. The gravelly and stony classes used were (1) go-nongravelly (0–15% grav-el), (2) g1-slightly gravelly (15–35% gravel), (3) g2-moderately gravelly (35–60% gravel), (4) g3-strongly gravelly (60–90% gravel), and (5) st5-strongly stony (>90% stones). The extent of area under each gravelliness/stoniness class association for the study area is given in Table 11.7.

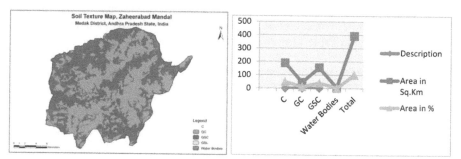

FIGURE 11.7   Area Extent of Soil Texture Units of Zaheerabad Mandal.

TABLE 11.7   Area Extent of Soil Texture Units of Zaheerabad Mandal

Mapping Unit	Description	Area in Sq. Km	Area in %
C	Clay	189.17	48.24
GC	Low	44.28	11.29
GSC	Very High	154.52	39.41
	Water Bodies	4.16	1.06
	Total	392.13	100.00

11.7   SLOPE

Slope refers to the inclination of the surface of the land. It is defined by gradient, shape and length, and is an integral part of any soil as a natural body. The length and gradient of slope influences soil formation and soil depth, which in turn affects land development and land use. Around 97 sq.km of area under mandal is level to nearly level slope, 220 sq.km of area is under very gently sloping lands and 72 sq.km is covered under gently sloping land (Table 11.8 and Fig. 11.8).

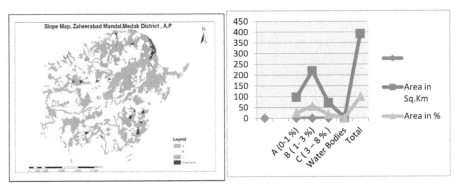

FIGURE 11.8 Area Extent of Slope Units of Zaheerabad Mandal.

TABLE 11.8 Area Extent of Slope Units of Zaheerabad Mandal

Slope Categories	Description	Area in Sq.Km	Area in %
A (0–1%)	Level to Nearly Level	97.07	24.76
B (1–3%)	Very Gently Sloping	219.20	55.90
C (3–8%)	Gently Sloping	71.70	18.28
	Water Bodies	4.16	1.06
	Total	392.13	100.00

11.8 LAND EVALUATION METHODS

Soil Map provides the user with information about the soil and landform conditions at any site of interest (Young, 1976). Soil maps are produced to suit the needs of users with widely different problems because they contain considerable detail to show basic soil differences (Klingebiel, 1966). Land evaluation is a tool for strategic land-use planning. It predicts land performance, both in terms of expected benefits from and constraints to productive land use, as well as the expected environmental degradation due to these uses (Rossiter, 1996). There is various land evaluation methods used for assessing the potential and productivity of soil for agricultural purposes. Some of them can be listed as below:

1. Land Capability Classification
2. Soil and Land Irrigability Classification

11.8.1   LAND CAPABILITY CLASSIFICATION

Land capability classification is an interpretative grouping of soils to show the capability of different soils to produce field crops or to be put to other alternative uses such as pasture, forestry, as habitat for wildlife, recreation, etc., on a sustained basis. It is based on inherent soil characteristics, external land features and other environmental factors that limit the use of the land (I.A.R.I., 1970). Eight land capability classes are identified. Soils suitable for agriculture are grouped under classes I to IV according their limitations for sustained agricultural production. Soils not suitable for agriculture are grouped under classes V to VIII for use for pasture, forestry, recreation purposes, quarrying and mining, and as habitat for wildlife. The land capability classes have subclasses to indicate the dominant limitation for agricultural use. Four kinds of limitations are recognized at the subclass level and denoted by "e" for problems caused by water and wind erosion, "w" for problems of drainage, wetness or overflow, "s" for soil limitations affecting plant growth like soil depth, heavy clay or sandy texture, gravelliness and stoniness, salinity or sodicity, etc., and "c" for climate limitation. The aerial extents are mentioned in the Table 11.9.

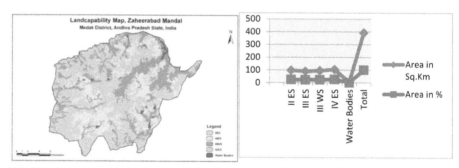

FIGURE 11.9   Area Extent of Land Capability Units of Zaheerabad Mandal.

TABLE 11.9   Area Extent of Land Capability Units of Zaheerabad Mandal

Land Capability Units	Area in Sq.Km	Area in %
II ES	98.25	25.05
III ES	90.46	23.07
III WS	94.96	24.31
IV ES	104.3	26.60
Water Bodies	4.16	1.06
Total	392.13	100.00

11.8.2 LAND IRRIGABILITY

Land irrigability classification is primarily concerned with predicting the behavior of soils under the greatly altered water regime brought about by introduction of irrigation. Land irrigability classification is an interpretative grouping of soil map units into soil irrigability classes based on the degree of limitations for sustained use under irrigation and on physical and socioeconomic factors (IARI, 1970) Soil irrigability classes are assigned without regard to the availability of irrigation water, water quality, land preparation costs, availability of drainage outfalls and other non soil related factors. Five soil irrigability classes are recognized. They are class A (none to slight soil limitations), class B (moderate soil limitations), class C (severe soil limitations), class D (very severe soil limitations) and class E (not suited for irrigation).

The suitability of land for irrigation depends on the quality and quantity of water, drainage requirements and other economic considerations in addition to the soil irrigability class. Lands suitable for irrigation are grouped under classes 1 to 4 according to their limitations. Lands not suitable for irrigation are grouped under classes 5 (temporarily classed as not suitable pending further investigations) and class 6 (permanently not suitable). Land irrigability classes have subclasses to indicate the dominant limitations for sustained use under irrigation. Three subclasses based on limitations are recognized and denoted by "s" for soil limitations such as heavy clay or sandy texture, soil depth, gravelliness and stoniness, "d" for drainage problems and "t" for limitations of topography.

The aerial extents of various land irrigability are mentioned in the Table 11.10.

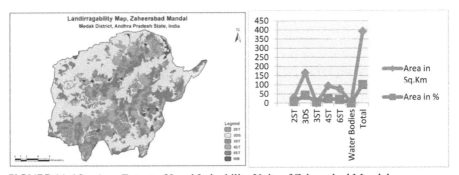

FIGURE 11.10   Area Extent of Land Irrigability Units of Zaheerabad Mandal.

TABLE 11.10   Area Extent Land Irrigability Units of Zaheerabad Mandal

Land Irrigability Units	Area in Sq.Km	Area in %
2ST	31.44	8.02
3DS	165.16	42.12
3ST	20 .36	5.20
4ST	94.06	23.98
6ST	76.95	19.62
Water Bodies	4.16	1.06
Total	392.13	100.00

11.9   FINALIZATION OF ACTION PLAN

To analyze and integrate spatial data using GIS, the thematic maps were digitized, coded and stored using GIS software packages. Intersection of various thematic maps was done progressively overlaying one layer over the other and by applying suitable decision criteria developed for that area. The final composite units bring out various types of homogeneous units/polygons. These would fit into the decision rules/recommendation arrived at. The draft action plan is discussed with the department officials, district administration, local research centers, NGOs working in the area and some progressive farmers for their feedback. Such feedback is critically evaluated considering the concept of sustainability for suitable modifications/improvements. Subsequently, the draft action plan is validated on the ground before being finalized.

11.10   RECOMMENDATIONS

The soil and land resource units of Zaheerabad Mandal were assessed and evaluated for their suitability for growing rice, sorghum, maize, sugarcane, cotton, green gram, black gram, bengal gram, red gram, groundnut, sunflower, soybean, safflower, castor, and guava which are the major crops grown in different parts of the mandal. The parametric approach of Sys (1985), which is a modified version of the FAO Framework for Land Evaluation (1976) was used for evaluating the land suitability. The Framework for Land Evaluation has recognized two orders, namely, order S-suitable for agriculture and order N-not suitable for agriculture. These orders are further subdivided into classes, subclasses and units. Order S-suitable for agriculture has three classes, highly suitable-S1, moderately suitable-S2 and marginally suitable-S3. Order N-not suitable for agriculture has two classes, N1-currently not suitable and N2-permanently not suitable for agriculture. The criteria assumed for differentiating into – sub classes are that, under a given management level specified to obtain optimum yield from a highly suitable land (Sl), the maximum reduction in yield successively may be in the order of about 25% for S2 and 50% for S3 classes. The suitability subclass reflects the kinds of limitations and indicate the kind of land improvements required within a class. The subclasses are indicated by the symbol using-lower case letters following the Arabic numeral (c-climate, e-erosion hazard, f-flood hazard, g-gravelliness, k-workability, I-topography, m-moisture availability, n-nutrient availability, p-crusting, r-rooting condition, t-texture, w-drainage, z-excess of salts/calcareousness). The suitability evaluation shows the areas that are highly suitable (class S1), moderately suitable (subclasses of S2), marginally suitable (subclasses of S3), currently not suitable; (class N1) and permanently not suitable (class N2) for each crop.

Land suitability for Rice: The land suitability assessment revealed that there are 4 suitability classes, 5 subclasses and 14 subclass associations with different kinds and degree of limitations. About 13358 ha (27.79%) area has been rated as moderately suitable in the mandal, 21614 ha (46.31%) area as Marginally and 14186 ha (43.60%) area as unsuitable (N2) for growing Rice crop.

Land suitability for Sorghum: The land suitability evaluation showed that there are 4 suitability' classes, 5 subclasses and 17 subclass associations with different kinds and degree of limitations. About 13358 ha (27.79% TGA) area has been rated as moderately suitable in the mandal, 21614 ha (46.31%) area as marginally and 14186 ha (43.60%) area as unsuitable (N2) for growing Sorghum crop.

Land suitability for Maize: The suitability evaluation indicated that there are 4 suitability classes, 6 subclasses and 16 associations of subclasses with different kinds and degree of limitations. About 10498 ha (21.01%) area has been rated as highly suitable (S1) in the mandal, 22940 ha (46.31%) area as moderately suitable (S2), 2862 (5.79%) area, as marginally suitable (S3) and 13322 (26.89%) area as unsuitable (N2) for growing maize crop.

Land Suitability for Sugarcane: The land suitability evaluation showed that there are 4 suitability classes, 5 subclasses and 15 subclass associations with different kinds and degree of limitations. About 10246 ha (20.64% TGA) area has been rated as highly suitable (S1) in the mandal, 15646 ha (31.53%) area as moderately suitable (S2), 16745 ha (33.75%) area as marginally suitable (S3) and 6985ha (14.07%) area as unsuitable (N2) for growing Rice crop.

The LRIS was designed for all the thematic layers generated in order to enable the users to work without help of traditional costly software's. The LRIS is a customized application which can be loaded in any system free of cost and be used by any person without formal training in GIS software's. It has some limited GIS query and analysis functions. The snaps shots of the LRIS for few thematic layers are presented in Fig. 11.11.

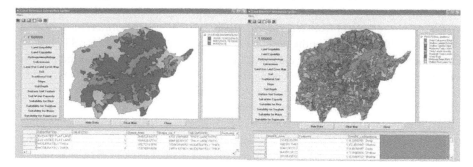

FIGURE 11.11   Snap Shots of the LRIs for Few Thematic Layers.

11.11   CONCLUSION

Earth observation satellites provide the vantage point and coverage necessary for studying our planet as an integrated physical and biological system. Land use planning involves the inventory of the land resources and taking stock of the present scenario. Land use planning does not only involve suggesting alternate land use but also should consider factors, which affect the other types of land use. The present study helped in the reconnaissance survey of the area as well as integrating the information to look at different scenarios in the landscape and plan for sustainable use of the land. The approach has given good insight into the areas potential for alternate land use. The action plan prepared using this approach shall help the administrators in taking decisions regarding resource use and mobilization of support for a change. The action plan not only serves as a guide but also as a blue print for natural resource management for sustainable development.

Compilation and collation of information of the area under study is the preliminary task in planning. The availability of remotely sensed data at high spatial and temporal resolutions has facilitated the planners to access natural resource information at a rate faster than never before. Land use map is the first map, which any planner would need to look at the extent of use to which the land is put. For purposes of planning at the district level the 1:50,000 maps generated using the IRS-P6 data were found to be suitable. Soil map and hydrogeomorphology map were the preliminary datasets generated from the satellite data. Information derived by integrating and analyzing the above factors were produced as derivative maps, which are the action plan maps.

KEYWORDS

- **Available Water Capacity**
- **Digital Elevation Model**
- **Geohydrology**
- **Geomorphology**
- **GIS Environment**
- **Graveliness**
- **Ground Water Potential Zones**
- **Land Evaluation Methods**
- **Land Irrigability**
- **Land Resource Information System**
- **Land Use Land Cover**
- **Relational Database Management System**
- **Soil Available-Water Capacity**
- **Soil Calcareousness**
- **Soil Depth**
- **Soil Erosion**
- **Stoniness**

REFERENCES

Aronoff, F. S. (1989). Geographic Information System; A Management Perspective, Ottawa, WDL Publications. 112–115.

Curran, P. J. (1988). Principles of Remote Sensing. 172–178.

Field Manual on Watershed Management- CRIDA, Hyderabad.

Hand Book of Statistics, Medak District 1997–1998.

Integrated Mission for Sustainable Development Manual, NRSA, Hyderabad.

Land Resource Management – Policy and Approach, National Discussion Paper of Agricultural Department. (2003).

Lille Sand J. M., Keiefer R. W. (1987). Remote Sensing and Image Interpretation; John Wiley & Sons. 82–86.

Remote Sensing Applications In Water Resource Development-Hand Book.

Soils of Deccan Plateau – Hand Book, Source: Soil Map of Andhra Pradesh, NBSS & LUP, Bangalore.

USGS Web Site.

CHAPTER 12

GEOMATICS FOR STUDYING THE TEMPORAL CHANGES IN LAND USE LAND COVER – A CASE STUDY OF SINDHUVALLIPURA WATERSHED, NANJANGUD TALUK, MYSORE DISTRICT, KARNATAKA

N. S. R. PRASAD[1], V. MADHAVARAO[2], R. R. HERMON[3], and P. KESAVA RAO[4]

CONTENTS

[1]Assistant Professor, CGARD, NIRD &PR
[2]Professor and Head CGARD, NIRD & PR
[3]Associate Professor CGARD, NIRD & PR
[4]Assistant Professor CGARD, NIRD & PR

ABSTRACT

The aim of this study was to detect spatial – temporal changes for land use and land cover using satellite imagery for 2000 and 2005 of Nanjanagud Taluk (Sindhuvalli-pura watershed), Mysore district of Karnataka and to gain the awareness on the fact that how geo-informatics is applied in the field of Rural Development. This change detection analysis was carried out by using the Geographic Information Software ERDAS IMAGINE 9.1 and Arc GIS 9.3. The images that were used for the analysis are the data products of IRS-Indian Remote Sensing Satellite and captured by LISS III and IV (Linear Imaging Self Scanning sensor). The images were geometrically corrected using UTM projection with WGS 84 Spheroid, followed by image processing operations unsupervised classification and post classification comparison change detection. Change detection analysis has been executed based on digital image classification and area statistic calculation.

12.1   INTRODUCTION

Soil, water, flora and fauna are the important land resources, which together influence the survival of human beings by supporting food production and providing a congenial living environment. Land resources are being exploited faster than they are renewed, as a result ecosystems are degraded, life support processes are threatened and biodiversity, being the key factor in maintaining biosphere resilience is decreasing at an alarming rate.

Monitoring and evaluation are vital components of watershed Planning and Development. The remote sensing technology has immense potential to meet the challenges of land resource management. The techniques of Remote Sensing and GIS systems provide a sound scientific knowledge to evaluate the watershed. The various techniques like supervised classification, Normalized Differential Vegetation Index and change matrices were used for the monitoring of changes in the watershed from the date of implementation.

12.2   OBJECTIVES

- To carryout natural resource inventory of land use/land cover using remote sensing data.
- To identify the temporal change in Land use/ Land cover for the study area.
- To assess the density and spatial distribution of Land use/ Land cover.

12.3   STUDY AREA

The study area (Fig. 12.1) comprises, Sindhuvallipura Watershed of Nanjanagud taluk, Mysore District, Karnataka, which spreads over an area of 3626.025 Ha. This area lies geographically between the 12°1'7.599" N to 12°1'46.973" N Latitude and 76°41'20.153" E to 76°39'19.999" E Longitude and falls in Survey of India Toposheet Nos. 57 D/12/SE and 57 D/12/SW scale 1:25,000.

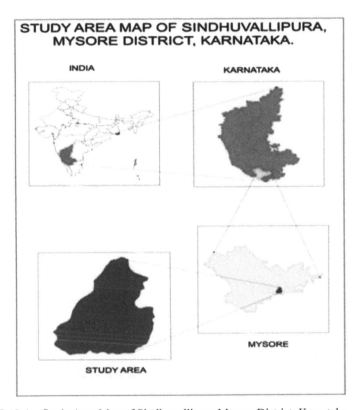

FIGURE 12.1   Study Area Map of Sindhuvallipura Mysore District, Karnataka.

12.4   METHODOLOGY

To achieve the objectives of the study, extent and density and type of vegetation and the vegetation growth, the integrated use of GIS and Remote Sensing and Digital Image Processing techniques were used for the study. The study was carried out specifically for the years, 2000 and 2005.

12.4.1   DATA ACQUIRED

For the study, satellite images of Nanjanagud Taluk (Sindhuvallipura) were acquired for two years 2000 and 2005. Toposheets bearing No 57 D/12 SE and 57 D/12 SW were collected from Survey of India, Hyderabad and IRS 1C, LISS III data of 23.5 mts and IRS P6, LISS IV 5.8 mts resolution procured from NRSC, Hyderabad. The procedure adopted is presented in Fig. 12.2.

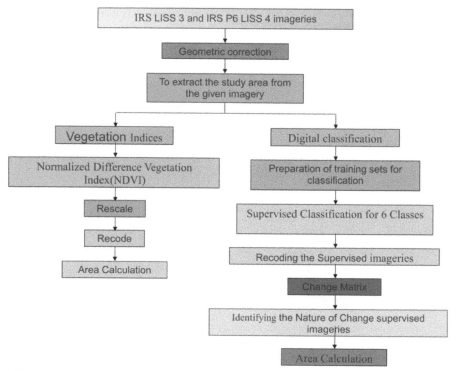

FIGURE 12.2   Flow Chart of Data Acquiring Procedure.

12.4.2   *LAND USE/LAND COVER CLASSIFICATION*

Thematic land classes can be derived digitally by grouping pixels that have similar spectral signatures from the measurements of individual bands throughout the spectrum. Supervised classification approach is used for forming classes that are similar in spectral reflectance.

12.5   LAND USE/LAND COVER MAP FOR 2000

The Land use/Land cover map for the year 2000 is prepared (Fig. 12.3). The analysis result shows that in the year 2000 (Table 12.1) the area falling under Crop 1 (Rice and Jowar) was 629.40 ha, Crop 2 (Ragi and Maize) was 430.78 ha. Area under

Water bodies is computed as 81.30ha, the area falling under Fallow Land 333.54 ha and the area is occupied by Land with Scrub 352.26 ha whereas 1804.63 ha area is occupied by land without scrub.

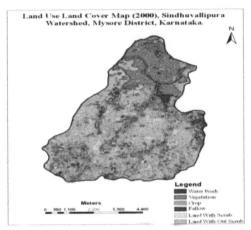

FIGURE 12.3   Land Use Land Cover Map (2000) Sindhuvallipura Watershed, Mysore District, Karnataka.

TABLE 12.1   Analyzed Results on Land Use and Land Cover Map for the year 2000

S. No	Class Name	Areas in Hectares	Areas in Percentage (%)
1.	Water Bodies	81.30	2.23
2.	Crop 1 (Rice and Jowar)	629.40	17.32
3.	Crop 2 (Ragi and Maize)	430.78	11.86
4.	Fallow Land	333.54	9.18
5.	Bounded Scrub	352.26	9.69
6.	Scrub Land	1804.63	49.67
	Total	3631.91	99.95

12.6   LAND USE/LAND COVER MAP FOR 2005

The Land use/Land cover map for the year 2000 is prepared (Fig. 12.4). The analysis result shows that in the year 2005 (Table 12.2) the area falling under Crop 1 (Rice and Jowar) was 1551.66 ha, Crop 2 (Ragi and Maize) was 329.22ha. Area under Water bodies is computed as 55.26 ha. 282.81ha area falling fallow and the area is occupied by Bounded Scrub (with scrub) 486.15ha, whereas 927.05ha area is occupied by land without scrub.

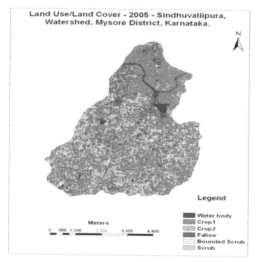

FIGURE 12.4   Land Use Land Cover (2005), Sindhuvallipura Watershed, Mysore District, Karnataka.

TABLE 12.2   Analyzed Results on Land Use and Land Cover Map for the year 2005

S.No	Class Name	Areas in Hectares	Areas in Percentage (%)
1	Water Bodies	55.26	1.52
2	Crop 1 (Rice and Jowar)	1551.66	42.72
3	Crop 2 (Ragi and Maize)	329.22	9.06
4	Fallow Land	282.81	7.77
5	Bounded Scrub	486.15	13.38
6	Scrub Land	927.05	25.52
	Total	3632.15	99.97

12.7   CHANGE DETECTION

Change detection analysis was carried out with the help of Change Detection Matrix (Fig. 12.5) provided with ERDAS imagine. By giving classified image of two different periods as input, the model automatically find out the area where changes are happened. For knowing changes happened in which type of land use classes statistical analysis (Table 12.3) were also carried out.

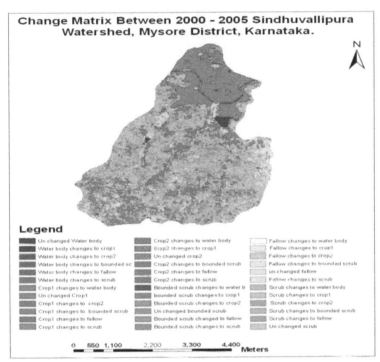

FIGURE 12.5 Change Detection Statistical Analysis of Land Use Land Cover between the Years (2000–2005).

TABLE 12.3 Change Detection Statistical Analysis of Land Use Land Cover Between the Years (2000–2005)

S. No	2000\2005	Water body	Crop 1 (Rice and Jowar)	Crop 2 (Ragi and Maize)	Fallow	Bounded scrub	Scrub	Total
1	Water Body	23.05	31.594	4.01	6.94	6.25	9.43	81.27
2	Crop 1 (Rice and Jowar)	8.14	460.17	79.47	32.94	12.82	35.49	629.03
3	Crop 2 (Ragi and Maize)	8.55	245.43	66.30	40.211	15.63	53.70	429.82

TABLE 12.3   *(Continued)*

S. No	2000\2005	Water body	Crop 1 (Rice and Jowar)	Crop 2 (Ragi and Maize)	Fallow	Bounded scrub	Scrub	Total
4	Fallow	1.52	87.91	22.80	80.23	27.82	132.00	352.28
5	Bounded Scrub	2.29	84.96	23.62	33.62	68.82	119.70	333.01
6	Scrub	10.00	641.30	132.95	292.13	151.42	576.60	1804.4
	Total	53.55	1551.36	329.15	486.07	282.76	923.74	3626.63

12.8   DETERMINATION OF VEGETATION DENSITY

Vegetation density analysis is usually carried out by calculating vegetation indices.

12.8.1   VEGETATION INDICES

A vegetation index is common spectral index that identifies the presence of chlorophyll. The index is composed of reflectance in the red spectral region (620 to 700 nm) and a portion (700 to 1100 nm) of the near-infrared (NIR) spectral region. Spectral satellite measurements in the red and infrared channels must be atmospherically corrected for interference from aerosols. Chlorophyll has a relative low reflectance in the red part of the electromagnetic spectrum (strong absorption) and relatively high reflectance in the near- infrared channels have been formulated.

The Normalized Difference Vegetation Index (NDVI) is a simple numerical indicator that can be used to assess whether the target being observed contains live green vegetation or not. Thus, NDVI was one of the most successful of many attempts to simply and quickly identify vegetated areas and their "condition," and it remains the most well-known and used index to detect live green plant canopies in multispectral remote sensing data.

The NDVI was carried out for the both the images, that is, 2000 (Fig. 12.6 and Table 12.4) and 2005 (Fig. 12.7 and Table 12.5) years.

TABLE 12.4 Normalized Difference Vegetation Index – 2000 – Sindhuvallipura, Watershed, Mysore District

S. No	Class Name	Area in Hectares	Area in Percentage (%)
1	Very Low	628.00	17.64
2	Low	934.57	26.25
3	Moderate	732.53	20.58
4	High	616.84	17.33
5	Very High	647.43	18.18
	Total	3559.37	99.98

TABLE 12.5 Normalized Difference Vegetation Index – 2005 – Sindhuvallipura, Watershed, Mysore District

S. No	Class Name	Area in Hectares	Area in Percentage (%)
1	Very Low	479.33	13.56
2	Low	694.55	19.66
3	Moderate	857.08	24.26
4	High	889.24	25.17
5	Very High	612.49	17.33
	Total	3532.69	99.98

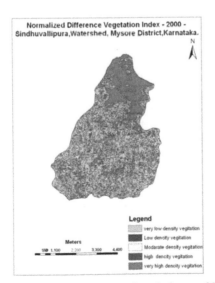

FIGURE 12.6 Normalized Difference Vegetation Index – 2000 – Sindhuvallipura, Watershed , Mysore District.

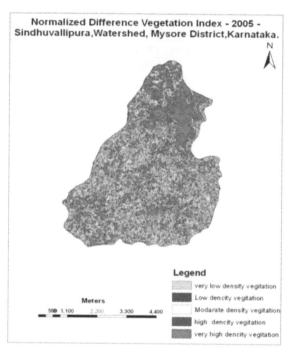

FIGURE 12.7 Normalized Difference Vegetation Index – 2005 – Sindhuvallipura, Watershed, Mysore District.

12.9   CONCLUSION

Result of Land use/Land Cover analysis of Sindhuvallipura, Nanjanagud taluk of Mysore district clearly indicate the different types of land cover in that area which are used for different purposes.

In 2000, land use/land cover shows the dominance of land without Scrub (49.67%), which covers an area of 1804.63 ha. Crop 1 (Rice and Jowar) (17.32%) is the second dominant class covers an area of 629.40 ha, followed by Crop 2 (Ragi and Maize) (11.86%) covers an area of 430.78 ha, Bounded Scrub (9.69%) with an area of 352.26 ha, Fallow Land (9.18%) with an area of 333.54 ha, water body (2.23%) covers less area, that is, 81.30 ha.

In 2005, Land use land cover shows the dominance of Crop 1 (Rice and Jowar) (42.72%) which covers area of 1551.66 ha, Land without Scrub (25.52%) is the second dominant class it covers area of 927 ha, Land with (Bounded) Scrub (13.38%) is the third dominant class it covers area of 486.15 ha, followed by Crop 2 (Ragi and Maize) (9.06%) covers area of 329.22 ha, Fallow land (7.77%) covers an area 282.81 ha and finally Water body (1.52%) covering less area of 55.26 ha.

In this study, Land Use/Land Cover map for the period of 2000 and 2005 was prepared which shows different results for that area (Table 12.6). The land without scrub is the dominant class and decreased from 2000–2005. The Crop 1 (Rice and Jowar) is the second dominant class it increased from 2000 to 2005, it was a positive change. The Crop 2 (Ragi and Maize) is the third dominant class it decreased from 2000 to 2005. The fallow land is decreased 2000–2005. The land with (Bounded) scrub increased from 2000–2005. The water body is decreased from 2000–2005.

TABLE 12.6 Change in Land Use Land Cover Results during the Period 2000–2005

Class Name	2000 Areas in Hectares	2005 Areas in Hectares	Change in Hactares
Water bodies	81.30	55.26	–26.04
Crop 1 (Rice and Jowar)	629.40	1551.66	+922.26
Crop 2 (Ragi and Maize)	430.78	329.22	–101.56
Fallow land	333.54	282.81	–50.73
Bounded scrub	352.26	486.15	ı133.89
Land without scrub	1804.63	927.05	–877.58
Total	3631.91	3632.15	

The results of change detection give us the idea about nature and causes of changes that happened in the period of 2000 to 2005. The change matrix analysis (Table 12.7) shows the changes in Croplands, water body, fallow land, land without scrub, and land with scrub.

TABLE 12.7 Change Detection in Nature and Cause of Changes During the Period of 2000–2005

Class Names	Areas in Hectares
Un changed Water body	56.97
Water body changes to Crop 1 (Rice and Jowar)	78.07
Water body changes to Crop 2 (Ragi and Maize)	9.92
Water body changes to fallow	17.16
Water body changes to bounded scrub	15.45
Water body changes to scrub	23.31
Crop 1 (Rice and Jowar) changes to water body	20.12
Un changed Crop 1 (Rice and Jowar)	1137.11

TABLE 12.7   *(Continued)*

Class Names	Areas in Hectares
Crop 1 (Rice and Jowar) changes to Crop 2 (Ragi and Maize)	196.38
Crop 1 (Rice and Jowar) changes to fallow	81.41
Crop 1 (Rice and Jowar) changes to bounded scrub	31.70
Crop 1 (Rice and Jowar) changes to scrub	87.72
Crop 2 (Ragi and Maize) changes to water body	21.14
Crop 2 (Ragi and Maize) changes to Crop 1 (Rice and Jowar)	606.49
Un changed Crop 2 (Ragi and Maize)	163.84
Crop 2 (Ragi and Maize) changes to fallow	99.36
Crop 2 (Ragi and Maize) changes to bounded scrub	38.63
Crop 2 (Ragi and Maize) changes to scrub	132.71
Fallow changes to water body	3.77
Fallow changes to Crop 1 (Rice and Jowar)	217.25
Fallow changes to Crop 2 (Ragi and Maize)	56.34
Un changed fallow	198.26
Fallow changes to bounded scrub	68.75
Fallow changes to scrub	326.19
Bounded scrub changes to water b	5.679
Bounded scrub changes to Crop 1 (Rice and Jowar)	209.959
Bounded scrub changes to Crop 2 (Ragi and Maize)	58.37
Bounded scrub changes to fallow	83.09
Unchanged Bounded scrub	170.07
Bounded scrub changes to scrub	295.79
Scrub changes to water body	24.71
Scrub changes to Crop 1 (Rice and Jowar)	1584.70
Scrub changes to Crop 2 (Ragi and Maize)	328.54
Scrub changes to fallow	721.88
Scrub changes to bounded scrub	374.16
Un changed scrub	1424.83

The results of NDVI mapping show the vegetation density of the area. In this map shows the very high density, high density, Moderate density and low density. So in the present study Crop 1 (Rice and Jowar) is having very high density, Crop 2 (Ragi and Maize) is the high density, Fallow land is having moderate density in that area and scrub is the low density in this area.

In 2000, Normalized difference vegetation index map shows the dominance of low density vegetation which covers area of, 26.25%, Moderate density vegetation is the second dominance class 20.58%, very high density is the third dominance class18.18%, followed by very low density vegetation 17.64%, finally high density vegetation covering the least area of 17.33% (Table 12.8).

In 2005, Normalized difference vegetation index map shows the dominance of high density vegetation which covers area of, 25.17%, Moderate density vegetation is the second dominance class 24.26%, low density vegetation is the third dominance class 19.66%, followed by very high density vegetation 17.33%, finally very low density vegetation covering the least area of 17.33%.

TABLE 12.8   Results of Normalized Difference Vegetation Index During the Period 2000 and 2005

Class Name	2000 Areas in Percentage (%)	2005 Areas in Percentage (%)
Very Low	17.64	13.56
Low	26.25	19.66
Moderate	20.58	24.26
High	17.33	25.17
Very High	18.18	17.33
Total	99.98	99.98

During the period 2000 to 2005 the moderate and high-density vegetations are increased. The very high density of vegetation was decreased by almost 1%. The remaining densities of vegetations are decreased.

In concluding, the satellite data can be effectively used in land use and land cover mapping, studying the land use changes and density of vegetation estimations.

KEYWORDS

- **Jowar**
- **Land Use/Land Cover Map**
- **Maize**
- **Near-Infrared**
- **Normalized Difference Vegetation Index**
- **Ragi**
- **Rice**
- **Vegetation Indices**

REFERENCES

Burrough, P. A. (1986). Principles of geographical information systems for land resources assessment. Oxford, University Press, 191p.

Jacob (1996). Planning and Management of Resource is an important area for the application of GIS.

Lillisand and Kiefer (1994). The Introduction of Advanced Image Classification techniques.

PART III

PLANNING FOR SUSTAINABLE AGRICULTURE, LAND USE IN DIFFERENT AGRO-ECOSYSTEMS

CHAPTER 13

RAINFED AGRO-ECONOMIC ZONES: AN APPROACH TO INTEGRATED LAND USE PLANNING FOR SUSTAINABLE RAINFED AGRICULTURE AND RURAL DEVELOPMENT

G. RAVINDRA CHARY[1]

CONTENTS

[1]Principal Scientist (Agronomy), Central Research Institute for Dry Land Agriculture, Hyderabad.

ABSTRACT

Rainfed agriculture accounting 55% of net cultivated area in India and remains main-stay for livelihoods of the majority of small and marginal farmers while contributing 40% to country's food basket. The biophysical and socioeconomic characteristics largely influence the rainfed agricultural land use. The impending climate change/variability is likely to accentuate the production related problems in rainfed regions. Both rainfed agriculture and rural development are land-based activities, hence an integrated land use planning therefore becomes necessary to achieve sustainable rainfed agriculture and rural development. For this purpose, Rainfed Agroeconomic Zones (RAEZs), as an area approach is discussed. Some of the learning experiences and concepts from the national agricultural research system that may be integral for R & D activities in RAEZs are also discussed. A sound Land Use Policy, separately for rainfed regions, at national level is necessary for creation of RAEZs which are contemplated to achieve much desired sustained development and enhance the livelihoods of the farmers in drought prone area and further, pave the way towards achieving "Second Green Revolution" from rainfed areas.

13.1   BACKGROUND

Rainfed agriculture in India is largely practiced in 76.7 m ha accounting 55% of the net cultivated area contributing 40% of food production with 85% coarse cereals, 83% pulses, 70% oil seeds and 55% rice. These regions are also home to 81% of rural poor, home to 66% of the total 458 million livestock. The details of rainfed area, state-wise are presented in Table 13.1.

TABLE 13.1   State-wise Net Sown Area, Net Irrigated Area, Net Rained Area and Percent Rainfed Area in India

State	Net Sown Area ('000 ha)	Net Irrigated Area ('000 ha)	Net Rainfed Area ('000 ha)	% Rainfed Area
Andaman and Nicobar $	9	2	7	78
Andhra Pradesh	9991	4214	5777	58
Arunachal Pradesh	212	56	156	74
Assam	2811	197	2614	93
Bihar*	5332	3394	1938	36
Chandigarh*	1	1	0	0
Chhattisgarh	4683	1323	3360	72
Dadra and Nagar Haveli*	20	4	16	80
Daman and Diu#	3	1	2	67

TABLE 13.1   *(Continued)*

State	Net Sown Area ('000 ha)	Net Irrigated Area ('000 ha)	Net Rainfed Area ('000 ha)	% Rainfed Area
Delhi @	23	22	1	4
Goa	132	29	103	78
Gujarat*	10,302	4336	5966	58
Haryana	3550	3069	481	14
Himachal Pradesh*	542	108	434	80
Jammu and Kashmir	735	317	418	57
Jharkhand	1250	102	1148	92
Karnataka	10,404	3390	7014	67
Kerala	2079	386	1693	81
Lakshadweep*	3	1	2	67
Madhya Pradesh	14,972	6892	8080	54
Manipur*	233	52	181	78
Meghalaya	283	62	221	78
Mizoram	123	10	113	92
Nagaland	361	73	288	80
Orissa	5574	2180	3394	61
Pondicherry	19	16	3	16
Punjab	4158	4073	85	2
Rajasthan	16,974	5850	11,124	66
Sikkim*	77	14	63	82
Tamil Nadu	4892	2864	2028	41
Tripura*	280	58	222	79
Uttar Pradesh*	16,589	13,457	3132	19
Uttarakhand	741	338	403	54
West Bengal*	5256	3112	2144	41
All India	140,022	63,256	76,766	55

Agricultural land use in rainfed regions, both temporally and spatially is influenced largely by biophysical characteristics viz. weather anomalies, soil quality, land degradation and socioeconomic parameters viz. farmers' resources, adaptive capacity of the farmers' to climate variability, farm inputs, market price, etc. Drought, particularly agricultural drought, either in season or within year is recurrent in one or the other part of the country in every year, therefore influencing agricultural land use thereby impacting production/ productivity of rained crops, and livelihoods of farmers in rainfed regions. Crop production in rainfed areas is generally affected by chronic (permanent) drought (South-west monsoon for the rainy season region and North-east monsoon for the Rabi region with low rainfall), ephemeral drought (occurring in early June, July; mid July August), terminal drought (September-early October periods during crop growth in South-west monsoon season), and apparent drought (South-west monsoon with high rainfall) (Fig. 13.1). There is a significant difference between chronic and other drought regions (Ravindra Chary et al., 2010).

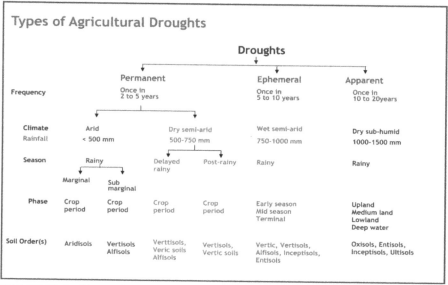

FIGURE 13.1   Types of Droughts in Rainfed Regions in India (Vittal et al. 2003).

In view of the vagaries in monsoon rainfall, agricultural drought can be anticipated to occur at any time within the crop growth period of the rainfed crops. In India, the drought regions can be classified into six regions (Fig. 13.2) based on the frequency of drought, climate, rainfall season, soil quality and soil orders (Vittal et al., 2003). In rainfed regions, the soils are generally degraded in quality and marginal in fertility. Alfisols (Red soils) and Vertisols (Black soils) of peninsular India and Aridisols (Desert soils) of extremely dry climates are the principal soil orders in dry areas, although Entisols (Alluvials) and Inceptisols (Alluvials) also occur, especially in topo-sequences. Vertisols, Alfisols, Entisols and associated soils are the major soil orders and are extensively under rainfed crops. About 30% of dryland area is covered by Alfisols and associated soils while 35% by Vertisols and associated soils having Vertic properties and 10% by Entisols of the alluvial soil regions. Rainfed farmers practice high diversity in cropping systems with livestock integration, which is an inbuilt risk management strategy.

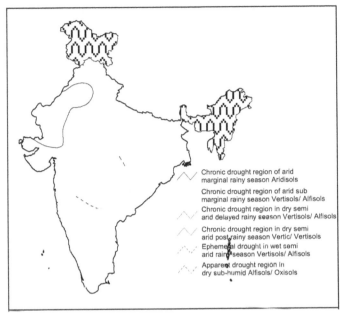

FIGURE 13.2   Spatial Distribution of Agricultural Drought Regions in India.

The cropping patterns have evolved based on the rainfall, length of the growing season and soil types. However, in recent decade, the land use diversification witnessed due to changed consumer preferences and market demand, farmers are now rapidly shifting to crops and cropping patterns, which are more remunerative. For example in Andhra Pradesh, the triggers for recent land use diversification in rainfed areas are: availability of new technologies and high returns (e.g., high yielding maize hybrids, Bt cotton, etc.), possibility of mechanization and good returns (300% increase in area of bold seeded chickpea on black soils in Prakasham and Kurnool district in one decade), non availability and rising labor costs and assured markets (e.g., Subabul and eucalyptus based farm forestry in 1.5 lakh ha. in Guntur, Prakasam, Khammam and Krishna districts in one decade on soils which are good for pulses and oilseeds, government programs (e.g., National Horticulture Mission triggering area under horticulture in several districts due to incentives like subsidy on drip irrigation and planting material, etc.), processing facilities (e.g., soybean area in Adilabad) and unfavorable weather/climate variability (e.g., sharp decline in short duration pulse area in Khammam, Krishna and Warangal districts due to rains during harvest).

There is now adequate evidence about the impending climate change and the consequences thereof (IPCC, 2007). Though climate change is global in its occurrence and consequences, it is the developing countries like India that face more adverse consequences. The significant aspect of climate change is the increase in the frequency of occurrence of extreme events such as droughts, floods and cyclones. All these changes will have adverse impacts on climate sensitive sectors such as agriculture and also on availability of water for agriculture (IPCC, 2007). More than seasonal rainfall, the distribution is more important for dryland crops grown during Kharif. Long dry spells have significant negative impact on fodder and grain production indirectly affecting the livestock production. Due to increased rainfall intensities land degradation is likely to increase in future. The expansion of rainfed agriculture as more and more regions become arid and semiarid, increased risk of crop failures and climate-related disasters and decreased yields are the important challenges that the changing climate will lead to. These will result in further deepening of poverty and food insecurity and loss of livelihoods in the rainfed regions. Land use practices also change air quality by altering emissions and changing the atmospheric conditions that affect reaction rates, transportation, and deposition. A microlevel study on climate variability and its impact on rainfed agricultural land use in Bhilwara district, Southern agroclimatic zone of Rajasthan indicated that the annual rainfall was decreasing by 46 mm in the last 40 years, reduction in number of rainy days in last 10 years, 21 years found to be arid and 23 years found to be semiarid in the last 45 years. In the last 15 years, shift in onset of monsoon observed, due to which the sowing window of maize shifted from 7th July to 20th July. This also impacted in high frequency of mid-season and terminal drought influencing change in varieties of Kharif and Rabi crops. There was change in cropping pattern for the last 20 years, that is, rainfed maize is replaced in the last 5 years by sesame, cluster bean and fodder sorghum. In the last 10 years, area under fodder sorghum/ black gram/sesame – fallow system has increased and area under maize-wheat has reduced and replaced in some areas with soybean – wheat/maize-mustard/taramira system. In Rabi, area under mustard and taramira increased.

The degraded lands particularly with water erosion account 73.27 m ha of arable lands (Anonymous, 2008). With depleting vegetative cover, the surface erosion problem is likely to accentuate further. This can further diminish soil fertility and will adversely affect native biodiversity, both above ground and below ground. Coupled with this, the demand for higher productivity would require greater nutrient input. Public investment in agriculture has progressively declined since two decades. Nor has private investment been able to fill this gap. As a result, agricultural growth is showing a declining trend. The private investment in rainfed areas is mainly on farm equipment and bore wells for groundwater exploitation. The productivity of crops in rainfed regions and income generated do not commensurate with the investment made by the farmers on adoption of improved technology when compared to irrigated regions. The socioeconomic parameters shown in Table 13.2 indicate an ample scope for investments in rainfed regions for overall development of the people, particularly the livelihoods of the farmers in rainfed regions.

TABLE 13.2   Relative Characteristics: Rainfed vis-à-vis Irrigated Regions

Parameter	Rainfed Regions	Irrigated Regions
Poverty ratio (%)	37	33
Proportion of agricultural labor (%)	30	28
Land productivity (Rs./ha)	5716	8017
Proportion of irrigated area(%)	15	48
Per capita consumption (kg/year) of:		
• Cereals	240	459
• Pulses	20	12
• Total food grains (kg/year)	260	471
Cooperative credit (Rs./ha)	816	1038
Bank Credit (Rs./ha)	1050	1650
Infrastructure development index	0.30	0.40
Number of predominant crops	>34	1*, 2**

Source: CRIDA Vision, 2030.
Note: * Rice-Rice in South India-wheat or Cotton-wheat in North India.
Source: (1) Source of Estimates of 17th to 4th rounds: NSO Report No.407; (2) GoI-NSSO 2006, p. 1.

There is also an immediate need of additional public investments to conserve natural resources for food security for now and in future as they are our National Capital. Post-harvest problems need more attention especially in rainfed areas, which have low management and adoption capacity and inadequate infrastructure. Rainfed areas are mostly monocropped and provide work for four to five months only. The interventions thus need to be designed in such a way that farmers are retained in farming round the year, may be through alternate livelihood options. One of the ways is agroforestry coupled with livestock on watershed mode besides other enterprises which could even be outside agriculture, meeting the local needs (skill enhancement of rural youth in processing and value addition techniques, etc.) and provide additional income and livelihood opportunities. In rainfed areas, the suitable interventions (as a basket of options to choose from) need to be propagated and also areas that are suitable for these specific interventions need to be delineated.

The demand for food grains in India is projected to be 308.5 m t by 2030 taking base year as 2004–2005 while the supply of food grains is projected as 265 m t based on projected population growth (0.95% per annum), thus leaving a gap of 43.1 m t. Diversification of land use is likely to be more in rainfed areas compared to irrigated and the contribution of rainfed agriculture would remain same at 44% to the total food grains.

13.2   INTEGRATED LAND USE PLANNING FOR SUSTAINABLE RAINFED AGRICULTURE AND RURAL DEVELOPMENT

Rainfed agriculture and allied enterprises, rural development activities like watershed management, land conservation/development, etc., are basically land based activities. Therefore, scientific land use planning at any level forms the basis to achieve sustainable rainfed agriculture and rural development. FAO (1995) defines Sustainable Agriculture and Rural Development (SARD) as a process which meets the criteria of: ensuring the basic nutritional requirements of present and future generations, qualitatively and quantitatively, are met while providing a number of other agricultural products; providing durable employment, sufficient income, and decent living and working conditions for all those engaged in agricultural production; maintaining and, where possible, enhances the productive capacity of the natural resource base as a whole, and the regenerative capacity of renewable resources, without disrupting the functioning of basic ecological cycles and natural balances, destroying the sociocultural attributes of rural communities, or causing contamination of the environment; and reducing the vulnerability of the agricultural sector to adverse natural and socioeconomic factors and other risks, and strengthens self-reliance. The elements and integrated activities for SARD include the policies, instruments, development plans, agrarian reforms, untraditional surveys, food quality and food security, data, monitoring, early warning systems at government level; development of local organization and capacity building for people's participation, training, extension at rural community level; coastal zones, watersheds, river basins, Agro-ecological zones at area level; farming systems, diversification to increase incomes, creation of rural industries, credit and marketing at production unit level and Improving nutrition and food quality, adjusting dietary patterns, product marketing at consumer level.

Land use planning appears to be a rational response to the above challenges. However, it is clear that attempts to meet such challenges are not keeping pace with the escalating severity of the problems.

- The very first step in land use planning has to be the explicit recognition of different goals of the various stakeholders, and definition of these goals in practical terms. It has become clear that outsiders cannot necessarily identify other people's priorities nor understand how best to meet them, so there must be direct negotiations between all interested parties to establish common goals, to trade-off between conflicting goals and, ultimately, decide between alternative courses of action.
- Empowerment of all groups of stakeholders to assume responsibility for land use and management.

- Incentives for land users to take the long view rather than opt for short-term exploitation of resources, and some means of restraining the feckless and greedy.
- Shrewd choices within the range of possible and attractive options require that information about all relevant factors is integrated at the point of decision.
- Sharing the benefits of common natural resource base may prove hard to negotiate. But the economic benefits can be significant if flexible transfers of a land and water are permitted within a well-constructed regulatory framework. These initiatives can only succeed if strong commitment is given to the participation of users in planning and investment decisions and the full and open sharing of economic and environmental information.
- Developed over millennia, the accumulated knowledge and biodiversity invoked needs to be preserved as a sound land use planning agricultural heritage system.
- Agrarian structure dominated by small holdings is no handicap for high agriculture production, provided requisite institutional support is available to the small farmers.
- To remain competitive and survive in the current economy, farmers must be insightful, innovative, and ready to make changes. In recent years, conventional wisdom has encouraged diversification with alternative enterprises and increased on-farm processing, packaging, and other means for adding value to raw products before they leave the producer's hands. While this makes good sense, making diversification and value-added strategies work can be challenging (Table 13.3).

TABLE 13.3 Issues and Functions Provided by Diversification in Rainfed Regions

Issue	Functions Provided by Diversification
Productivity and stability	Increased yield, reduce intra seasonal variation and improved stability through diverse components viz. crop, tree, plant and animal
High risk and high cost	Risk and cost minimization through yield and income from annual and perennial mixtures
Unabated land degradation	Minimization of kinds, effect and extent of land degradation by appropriate land care through alternate land use systems
Inadequate employment	Staggered employment round the year
Low profitability	High income generation from various components
Poor energy management	Energy efficient implements and livestock

Inhibitor Externalities in Scientific Land Use Planning With Globalization, trade in agricultural commodities was on progressively higher scale, but gains were not substantial due to unhealthy approach of the developed countries.

- Non-farm enterprises should be inhibited practicing on the agricultural land to avoid pushing agriculture into more marginalized lands.
- Technology and institutional changes designed to preserve the natural resources will have to keep the small farmer at the center.
- Often unreliable water deliveries forced users to overexploit groundwater as a source of environmental degradation.
- Management focused solely on crops will become unsustainable in economic and environmental terms
- Development itself is driven by economic and social factors such as demand and supply of products, markets, social and economic organization, and infrastructure.

13.3   RAINFED AGROECONOMIC ZONES (RAEZS) – THE APPROACH

The existing domain/area specific land based approaches to address rainfed agriculture and rural development like watershed development programs, etc., are either sectoral in approach or lacks sound scientific basis for land use. Further, the earlier developmental and relief measures focused through Drought Prone Area Program (DPAP) and Desert Development Program (DDP) in drought prone areas, which are primarily located in core rainfed agriculture regions, by the central and state governments could address only some of the drought impacting issues. Besides these, several other approaches were also adopted for delineation of rainfed areas for research and development. The Planning Commission identified 15 agro-climatic regions (ACRs) on geographical basis for development purpose and these ACRs cut across the states. ICAR, in 1979, under National Agricultural Research Project (NARP), delineated country into 127 agro-climatic zones (ACZs) based on soils, climate, topography, crops, vegetation, etc. However, each ACZ covers 2–4 districts and spread over an area of 40 to 50 thousand sq.km. NBSSLUP delineated country into 20 Agro-ecological regions (AERs) and 60 Agro-ecological subregions based on bioclimate (rainfall, temperature, vegetation and PET), length of growing period and soil scape (Sehgal et al., 1992), Further, Velayutham et al. (1999) have suggested agro-ecological subregions with details of the agro-ecological settings and the land use potentials and constraints of each subregion, can help to develop alternate land use plan scenarios, etc., for planning and development. Under National Agriculture Technology Project (NATP) the concept of Production System was introduced by ICAR (NATP, 2004) for research and extension purpose. In this approach, the rainfed agroecosystem was subdivided into five homogeneous production systems viz. Rainfed Rice based Production System, Nutritious Cereals(coarse cereals) based

Production System, Oilseeds based Production System, Pulses based Production System and Cotton based Production System. Presently, the focused issues in rainfed areas are resource conservation and management, increased productivity and profitability, making rainfed agriculture a viable profession and improving the livelihoods of the farmers/people in rainfed areas. Further, the major issues/ programs for SARD related to Integrated Land Use Planning are:

- climate variability/change;
- land degradation;
- agriculture policy review, planning and integrated programming in the light of the multifunctional aspect of agriculture, particularly with regard to food security and sustainable development;
- improving farm production and farming systems through diversification of farm and nonfarm employment and infrastructure development;
- land resource planning information and education for agriculture;
- land conservation and rehabilitation;
- water for sustainable agriculture and sustainable rural development;
- diversification for income and employment generation;
- reducing the vulnerability of rural communities to risks and uncertainties;
- enabling rural communities for better resilience;
- traditionally, developmental programs were imposed in a top–down approach (this approach has been attempted at state, district and village level);
- data requirements would vary from broad physical resources at state level to more detailed agro-ecological, social and economic resources at microlevel.

In light of the rainfed agriculture scenario, SARD issues, land use planning issues discussed earlier, to achieve twin goals of sustainable rained agriculture and rural development, it calls for an entirely a new target domain approach. This could be by delineating core 'Rainfed Agro-Economic Zones' in a district or part of a region in a state (Ravindra Chary et al., 2008). The important criteria for delineation of these RAEZs could be a predominantly rainfed region with predominant rainfed production system (having 80% net rainfed area), source and percentage net irrigated area (<25% net irrigated area by all sources) and with probability of recurrence of any type of drought, that is, permanent, ephemeral and apparent drought. Here, the rainfed farmers' livelihoods improvement and sustaining the land resources would be focal, wherein all the interventions related to land resource management, soil and water conservation, rainwater management (in situ and ex situ), management of rainfed crops for higher land and water productivity, real-time contingency and compensatory crop planning, farm mechanization for precision and timely agricultural operations will be implemented specific to land resources, marketing opportunities both on-farm (at farm gate) and off-farm, warehouses and cold storage facilities, post harvest value addition with value chain facilities for higher profitability leading to enhanced livelihoods. Instead of individual and piecemeal interventions the entire rainfed production system will be targeted to develop as a RAEZs, which would act as hubs of rainfed agriculture development. Tools like crop or weather insurance, where the insurance product may have an inbuilt condition that a given product is applicable only if a particular crop or commodity is grown in RAEZs where in the government can subsidize part of the premium for farmers who adopt rational water use. New opportunities are also arising in the area of Clean Development Mechanism (CDM) and carbon credits, which can be implemented in RAEZs where farmers can be compensated for adopting soil, water and energy conservation practices and carbon emission reduction practices, which also contribute to drought amelioration and sustainable productivity on a long-term basis, but relatively lower returns on short-term. For example, Malwa Plateau zone in Madhya Pradesh could be a RAEZ for soybean based rained production system, Anantapur district in Andhra Pradesh could be a RAEZ for groundnut based production system and so on. Further, these RAEZs could simultaneously address nonarable lands and common lands for suitable and multiple land uses for ecosystem services and other rural development activities. RAEZs akin to agriexport zones have greater scope in hill and tribal districts of the country with focus on organic farming zones, for example Kandhmal district in Odisha, a truly organic and agriexport zone for rainfed turmeric well supported by KASAM (a farmers' association) and local government.

13.4   SOME LEARNING EXPERIENCES, RESEARCH OUTCOMES, CONCEPTS INTEGRAL FOR VARIOUS R&D ACTIVITIES IN RAEZS

(a) Learning Experiences from NATP-Mission Mode Project on Land Use Planning for Management of Agricultural Resources in Rainfed Agroecosystem

NATP-Mission Mode Project on Land Use Planning for Management of Agricultural Resources in Rainfed Agroecosystem, was undertaken in 16 microwatersheds during 2002–2005 in an area of 5258 ha across arid, semiarid and subhumid agroecosubregions with 13 microwatersheds adopted by 13 network centers of All India Coordinated Research Project for Dryland Agriculture (AICRPDA), 2 regional centers of National Bureau of Soil Survey and Land Use Planning, and one microwatershed by CRIDA as Lead center. The project was implemented in a bottom-up and participatory mode, detailed soil resource characterization was done and delineated 132 soil-sub groups under five major soil orders (Entisols, Inceptisol, Alfisols, Vertisols and Aridsols) in 5258 ha across 16 microwatersheds located in arid, semiarid and subhumid agro-ecological-subregions. The PRA and socioeconomic inventory of 1763 households in 16 microwatersheds indicated 14 biophysical, 9 socioeconomic, 9 production, 14 infrastructure and 13 technical constraints for sustaining the land productivity under rainfed situation. The traditional land use across dominant crop based rainfed production systems viz. groundnut, pearl millet, Rabi sorghum, soybean, cotton, upland rice, finger millet, maize and Kharif sorghum. While the traditional land use in relation to soil-site suitability in varying soil types in rainfed agroecosystems indicated more incorrect agricultural land use practiced by farmers compared to correct land use (Table 13.4).

TABLE 13.4   Traditional Land Use in Major Soils Regions in Rainfed Areas

Major Soils Region	Land Use Practiced by Farmers	
	Correct #(%)	Incorrect #(%)
Black soils	81(45)	98(55)
Red soils	21(40)	31(60)
Others	20(71)	8(29)
Total	122(47)	137(53)

#: Number of locations.
Source: NATP-MM-LUP- Rainfed Final Report (2006).

The microlevel participatory land use planning process included PRA, farmers' choice and soil-site suitability evaluation for suitability of crops. About 932 on-farm demonstrations over 603 ha at 1294 sites on 132 soil subgroups on varying topo-sequences were conducted in 16 microwatersheds. This provided much needed land use diversification from the traditional rainfed land utilization. These interventions for successful participatory land use planning included critical dry land practices, crops, cropping systems and alternate land use systems. Microlevel variations of soils (phases of soil series) and management practices on a topo-sequence are the prime factors influencing land productivity. The promising interventions/practices included cotton+groundnut (1:1) groundnut(GG-20)+blackgram (T-9) and cotton(H8) + sesame intercropping system in deep swell-shrink black soils (Typic Haplusterts) at Vagudad watershed; newly introduced Macuna utilis a medicinal crop during drought in deep red soils (Rhodic Paleustalfs) at Amanishivapurkere watershed; pearlmillet (GHB-235)+greengram (GM-4) (3:1) inter cropping in medium deep sandy loam soils (Typic inceptisols), improved hybrid castor (GCH-5) with sulfur application@25 kg/ha) in shallow sandy loam soils (Typic Ustorhents – and line sowing of clusterbean (GC-2) (45x10 cm) at Saleempura watershed, SK Nagar; in >100 cm deep black soils, chickpea cv. A-1 and groundnut +castor (7:1) intercropping, in <50 cm deep soil, coriander (improved variety –DWD-3), chickpea cv. A-1 at Gadehotur watershed, Bellary; coriander+chickpea+senna relay intercropping as contingency crop planning to cotton crop in medium deep swell-shrink black soils (Typic Haplusterts) at Chidambarapuram watershed, Kovilpatti; soybean pigeonpea as sole and as intercrop in deep swell-shrink black soils (Typic Haplusterts) at Khuj watershed, Rewa; desi cotton in medium deep black soils (Typic Ustochrepts), sesame cv.AKT-64 (28q/ha), castor dwarf variety (AKKC-1), cucurbits cv. Ankur Himangi (41.8 q/ha) and Matki (local pulse,1.95 q/ha) in deep black soils (Vertic Inceptisols) at Varkhed watershed, Akola; newly introduced crops blackgram, safflower, vegetables and agrohortisystem (citrus+soybean, citrus+wheat) in deep swell-shrink black soils (Typic Haplusterts) at Jaitpura watershed, Indore; sorghum cv.CSH-9, BJH-117, soybean(JS-335) in deep black soils (Vertic Ustochrepts) and introduction of LRK-516 cotton in shallow-skeletal black soils (Lithic Ustorthents) and NHH-44 in calcareous medium deep black soils (Typic Inceptisols) at Panubali watershed, Nagpur; introduction of safflower with compartmental bunding, pigeon pea cv. ICPL-87 in intercropping of sunflower+ pigeon pea (2:1) and sequence cropping of greengram – Rabi sorghum in deep swell-shrink black soils (Typic Haplusterts) at Sarole watershed, Solapur; and diversification to maize, sorghum and pigeonpea based intercropping system in Entisols (shallow sandy clay loams) with 1 to 3% slope under well drained uplands and medium lands, conservation tillage for in-situ moisture conservation in all soil types, crop substitution with chickpea in place of wheat in Udic Eutrudepts at Ballipur watershed, Varanasi, the Land productivity increase was 30 to 50% and in few cases more than double.

With the experiences from the project outcomes Ravindra Chary et al. (2005) developed a conceptual approach for delineation of Soil Conservation Units (SQUs), Soil Quality Units (SQUs) and Land Management Units (LMUs) from the soil resource information at cadastral level in a microwatershed as these would likely to help in land resource management since these units are homogeneous in a theme and have a wider application. A resilient, less risk prone farming system based on the land requirements and farmers' capacities could be developed to mitigate both in-season and impending drought, to address the unabated land degradation and imminent climate change. The SCUs form the basis as to prioritize and implement suitable interventions and programs/schemes related to soil and water conservation and has a wider scope to converge with programs like National Rural Employment Guarantee Scheme (NREGS), other various land management/conservations/development programs in a watershed mode by Ministry of Rural Development either at Central or at State level. The SQUs are to address soil resilience and improve soil organic carbon, problem soils amelioration and wastelands treatment and linked to various schemes and programs in operation like Rashtriya Krishi Vikas Yojana (RKVY), National Horticultural Mission (NHM), Boochetana (Karnataka), Network planning (NABARD), etc. Further, SCUs and SQUs are merged in GIS environment to delineate land parcels into homogenous Land Management Units with farm boundaries. LMUs would be operationalized ultimately at farm level for taking land resource management decisions on arable, nonarable and common lands for crop intensification/diversification, alternate land use systems like agroforestry, agrohorticulture, ecorestoration, etc. Cadastral level Rainfed land use planning modules should be based on these units for risk minimization, enhanced land/water productivity, income, drought proofing, land resource conservation and finally for enhancing livelihoods of the rained farmers. An example of delineation of these units for the Kaulagi watershed, Bijapur district, Karnataka is shown in Figs. 13.3–13.6.

FIGURE 13.3   Soil Resource Map – Kaulagi Watershed, Bijapur district, Karnataka.

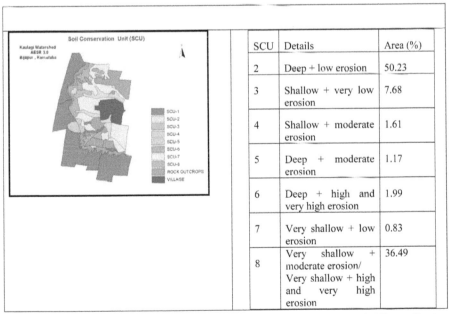

SCU	Details	Area (%)
2	Deep + low erosion	50.23
3	Shallow + very low erosion	7.68
4	Shallow + moderate erosion	1.61
5	Deep + moderate erosion	1.17
6	Deep + high and very high erosion	1.99
7	Very shallow + low erosion	0.83
8	Very shallow + moderate erosion/ Very shallow + high and very high erosion	36.49

FIGURE 13.4   Soil Conservation Units (SCUs) in the Kaulagi Watershed, Bijapur district, Karnataka.

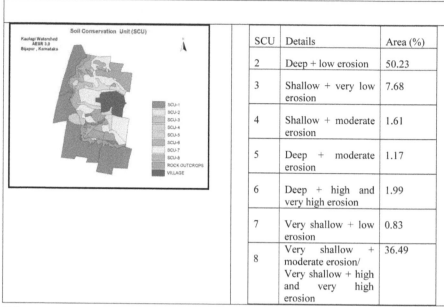

SCU	Details	Area (%)
2	Deep + low erosion	50.23
3	Shallow + very low erosion	7.68
4	Shallow + moderate erosion	1.61
5	Deep + moderate erosion	1.17
6	Deep + high and very high erosion	1.99
7	Very shallow + low erosion	0.83
8	Very shallow + moderate erosion/ Very shallow + high and very high erosion	36.49

FIGURE 13.5   Soil Quality Units (SQUs) in the Kaulagi Watershed, Bijapur District, Karnataka.

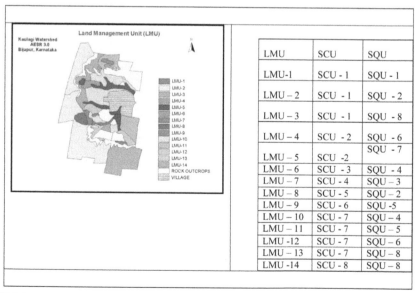

LMU	SCU	SQU
LMU-1	SCU - 1	SQU - 1
LMU – 2	SCU - 1	SQU - 2
LMU – 3	SCU - 1	SQU - 8
LMU – 4	SCU - 2	SQU - 6
LMU – 5	SCU -2	SQU - 7
LMU – 6	SCU - 3	SQU - 4
LMU – 7	SCU - 4	SQU – 3
LMU – 8	SCU - 5	SQU – 2
LMU – 9	SCU - 6	SQU -5
LMU – 10	SCU - 7	SQU – 4
LMU – 11	SCU - 7	SQU – 5
LMU -12	SCU - 7	SQU – 6
LMU – 13	SCU - 7	SQU – 8
LMU -14	SCU - 8	SQU – 8

FIGURE 13.6   Land Management Units (LMUs) in the Kaulagi watershed, Bijapur District, Karnataka.

A Framework of Rainfed Land Use Plan for Participatory Self-Sustaining Viable Bio-diverse Mixed Farming System with activities, relevant partners, strategies and outcomes at different scales was suggested (Ravindra Chary et al., 2004) (Fig. 13.7), which may be integral to RAEZ concept for participatory and use planning at cadastral level.

AICRPDA–All India Coordinated Research Project for Dryland Agriculture; AICRPAM–All India Coordinated Research Project on Agro-Meteorology; CRIDA–Central Research Institute for Dryland Agriculture; SAU–State Agricultural University; NBSS&LUP–National Bureau of Soil Survey and Land Use Planning; IGFRI–Indian Grasslands and Fodder Research Institute; NRCAF–National Research Centre for Agroforestry; NARES–National Agriculture Research and Extension System (ICAR, SAUs, Private corporate research); NGOs–Nongovernmental Organizations.

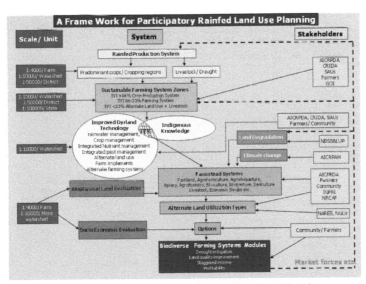

FIGURE 13.7   A Framework for Participatory Rained Land Use Planning.

(b) Some Proven Interventions for Drought Management in Rainfed Areas – Experiences of AICRPDA

(i) Permanent Drought Amelioration

In drought prone regions, rainwater management is the foremost activity to be undertaken. The core strategies of rainwater conservation are based on improving the water availability to the crops and increasing the groundwater recharge. The approaches are on building in situ moisture reserves to tide over the recurring drought spells, disallowing subsequent loss of soil-profile stored moisture, permitting safe runoff disposal, its collection above or below ground, and tactical recycling of harvested runoff. Applicability of techniques so developed varied with the soil hydraulic characteristics. Salient findings and/or interventions in the theme area of rainwater management, which emerged as research outcomes from the network centers of All India Coordinated Research Project for Dryland Agriculture (AICRPDA) could be effective for drought management (AICRPDA, 2003) and shown in Fig. 13.8. Integrated watershed management is the key to conservation and efficient utilization of vital natural resources viz. 'soil' and 'water,' particularly in rainfed agriculture where water is the foremost limiting factor for agricultural production. The prioritized steps involved in resource conservation are the use of practices based on the existing traditional systems.

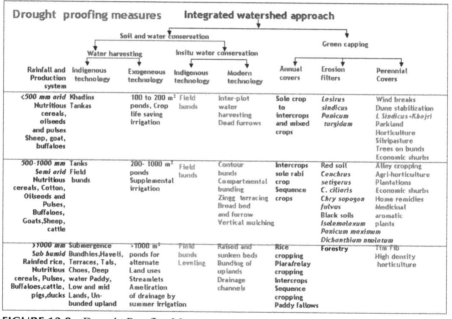

FIGURE 13.8   Drought Proofing Measures.

Location specific crop/contingency plans for mitigating the effects of drought have been recommended by various research stations of state agricultural universities and also by the network centers of AICRPDA (Subba Reddy et al., 2008), applicable to their respective domains and an example each for chronic, apparent and ephemeral drought situations (Vittal et al., 2003) is given in Table 13.5.

TABLE 13.5 Crop and Contingency Plans for Different Drought Types Based on Climate and Soil Orders

Recommendation Domain	Crop Plan			
	June second fortnight to July first fortnight)	Second fortnight of July	First fortnight of August	Second fortnight of August
1.0. Chronic Drought				
1.1 Chronic drought in arid region in marginal soils with rainfall < 500 mm				
Hisar, Bhiwani, Sirsa, Mahendragarh, Gurgaon and part of Rohtak district of Haryana in marginal hot typic arid aridisols	Pearlmillet Cluster bean Greengram	Short duration crops Pearlmillet (HHB-67) Greengram (Asha) Blackgram (T-9) Cowpea (Charodi) Clusterbeans (HG 365)	Nursery of Pearlmillet (HHB 67) nursery may be kept ready for transplanting. Crops Fodder	Toria with supplemental irrigation Rains received after mid of August may be conserved for rabi sowing.
1.2 Chronic drought in arid region in sub marginal soils with rainfall 500–750 mm				
Rajkot, Surendranagar, Jamnagar, parts of Junagarh, Bhavanagar and Amreli districts of Gujarat in hot arid Vertisols	Sorghum, Pearlmillet, Groundnut, Castor, Cotton	Erect groundnut (GG-2, GG-5, GG-7) Sesame (G. Til-1, G. Til-2);Pearlmillet (GHB-235 GHB-316, GHB-558) Greengram (K-851, GM-4); Blackgram (T-9;) Pigeonpea (ICPL-87, GT-101)	Blackgram (T-9), Forage maize/ sorghum (Gundri, GFS-5), Castor (GAUCH-1) Sesame (Purva-1)	Forage maize Sorghum (Gundri, GFS-5) Sesame (Purva-1)
1.3 Chronic drought in dry semi – arid region with rainfall 500–750 mm (late rainy season to post-rainy season)				
Madurai, Ramanthapuram and Tirunalveli districts of Tamil Nadu in hot moist semi arid alfisols/ Vertisols	Onset of monsoon is from September			

TABLE 13.5   *(Continued)*

1.4 Chronic drought in dry semi-arid region with rainfall 500–750 mm (post-rainy season)

Bijapur and Gulbarga districts, parts of Belgaum, Linguagur of Raichur district of Karnataka and southern parts of Maharashtra in hot semi-arid Vertisols	Greengram Blackgram Pigeonpea Pearlmillt	Groundnut (spreading) Hybrid pearlmillet Sunflower and Setaria in kharif areas pure pigeonpea/ cowpea/ horsegram in light soils	Cotton in middle of August. Early sowing of cotton is advantageous Grow herbaceum cottons in place of hirsutums Sunflower Pigeonpea, castor and setaria in light soils. Pigeonpea in medium to deep black soils	In medium to deep black soils, on contour bunds Castor Relay cotton in groundnut in medium black soils.

2.0. Ephemeral Drought

2.1 Ephemeral drought in wet semi – arid region with rainfall 500–750 mm (rainy season)

Bhilwara, Tonk, Dungarpur, Ajmer districts and parts of Bundi, Chittaurgarh, Rajasamand in hot dry semi-arid inceptisols/ aridisols	Pigeonpea Cowpea	Sesame (RT-46) Greengram (K-851 and RMG-62);Sorghum (Fodder);Cowpea (Fodder) (Raj Chari-1 and 2, C-152) No sowing of cereals Only short duration pulses and oil seeds or fodder crops should be sown. Soil mulching and interculture to conserve soil moisture is beneficial.	Sesame (RT-125) Greengram (RMG-62) Sorghum (fodder) (Raj Chari-1) (single cut) Soil water conservation measures for in situ management and runoff harvest for recycling during later part of crop season.	Sorghum (Fodder) (Raj Chari –1) Toria (TL-15) Tarmaira (T-27) Rain received after first fortnight of August should be conserved for early rabi seeding of toria/ taramira during first week of September. Any heavy downpour occurs, harvest the water for pre-sowing supplemental irrigation to rabi crops.

TABLE 13.5   *(Continued)*

3.0. Apparent Drought

3.1 Apparent drought in dry sub humid region with rainfall 1000–1500mm (rainy season)

Uplands and medium lands of Balasore, Cuttack, Puri and Ganjam districts of Orissa in hot moist sub-humid inceptisols	Pigeonpea, Mesta, Maize, Groundnut, Fingermillet, Rice, Sorghum, Cowpea, Blackgram, Greengram	Upland Blackgram Setaria (Pant -30);Green-gram (PDM54/K 851);Sesame (Uma or local) ; Early Pigeonpea (UPAS 120/ICPL-87); Planting of short duration vegetables as rad-ish (Pusa Chetki), okra, cowpea (SEB-2/SEB 1) and clusterbeans.	Upland Sowing of niger, black-gram, sesame, greengram, Planting of vegetables as radish, beans, cowpea, Early Pigeon-pea (ICPL-87/ UPAS-120)/W Medium and shallow submerged lowland Direct line sowing of extra early rice (Heera, Vandana, Kalinga-III, Z HU 11–26, Rudra, Sankar and Jaldi-5).	Upland Horsegram; Sesame Niger; Cowpea

13.5   PRODUCTIVE FARMING SYSTEMS: A3 X3 MATRIX APPROACH

Productive farming systems are identified for drought prone regions based on annual average rainfall, land capability and soil order (Vittal and Ravindra Chary, 2007). Land use based farmstead plan state-of-the-art based agroforestry models linked to livestock and watershed management for soil and water conservation including water harvesting. Some of the subjects are hedge fencing, multipurpose tree species, bush farming, cereals/millets. Pulse/oilseeds/cotton, parkland horticulture, olericulture, floriculture cum IPM, home remedies, water harvesting, livestock, poultry, fisheries, apiary, etc., are some of the models suggested into higher value agricultural crops (medicinal, aromatic, dye yielding crops, etc.), and nonfarm activities like value addition to agricultural products offer good scope for sustained increase in per capita income. Part of the farmstead could also be used for generating seed spices. For the development of commons, these may be divided into small plots of 5–10 ha and can be put on long lease of about 19 years to the user groups. The combination of systems such as fruit trees, silvipastures, multipurpose trees, or even pastures may be adopted on commons. Maximum number of trees per hectare may be limited by quantum of annual rainfall (product of rainfall in m and area in m^2) divided by volume of water one full grown tree transpires annually (a product of canopy area, surface area in m^2, and potential evapotranspiration in m per annum). Improved variety or new plant species suitable for the ecosystem and rejuvenation of social fencing of improved plant species may be attempted. In tree farming, the general cleanliness of the area is lost thus, encouraging new diseases and pests. Therefore, it is important to carry out weeding and form basins for the trees and furrows for in situ rainwater harvesting in the case of shrubs, grasses and fodder legumes. Alternate land use systems can provide stability of production and income through alley cropping, agroforestry, medicinal plants, horticulture, etc., in place of annual crops on marginal lands. They help in better soil and water conservation. Some promising alternate land use systems are shrub farming (henna, curry leaf, Jatropha, Karanji), horticulture systems (guava, ber, pomegranate, custard apple), and fodder species (Leucaena, glyricidia, Stylo santhes). At present, farmers' cropping systems are neither providing stability nor balancing input-output nutrient relations. Markets in the developed world are evincing a keen interest in organic farming. Rainfed agriculture involves fewer external inputs, and provides good opportunities to adopt organic farming in hill and tribal districts of the country. It must be remembered that the objective of diversification is to distribute risk, not to increase it through poorly conceived undertakings. Success or failure can depend on a number of factors; one of these is good information. Before plunging into new, costly ventures, the following advice is worth heeding. Anticipated benefits of crop diversification are: alternative crops may enhance profitability, diversified rotations can reduce pests, labor may be spread out more evenly, different planting and harvesting times can reduce risks from weather, new crops can be renewable resources of high value products and soil health is taken care of and land degradation is minimized. The emphasis here in agro-ecological analysis is on the processes and balance of resource supply and capture, and on the competitive and complementary relationships between the planned and unplanned (associated) biodiversity. Diversification strategies should be based on low external input strategies (Vittal et al., 2007) (Fig. 13.9).

Productive Farming Systems

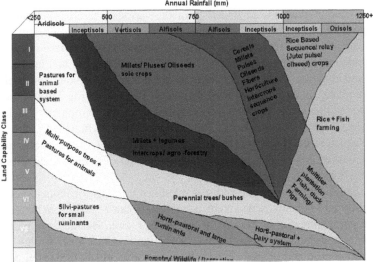

FIGURE 13.9   Productive Farming Systems Matrix in Rainfed Regions.

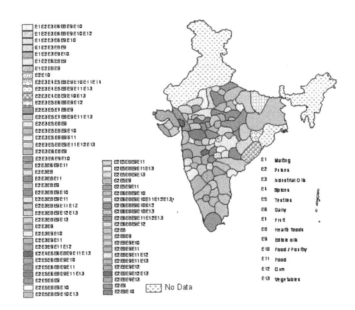

FIGURE 13.10   Agro-Industries with Potential Crop Diversification.

13.6   AGRO-ENTREPRENEURSHIP WITH POTENTIAL CROP DIVERSIFICATION IN RAINFED AREAS

From the recommendations emanated from the various intercropping and sequential cropping studies from different AICRPDA network centers, efficient cropping systems (intercropping and double cropping) were identified. These systems were recommended to similar agroecoregions. Based on the existing dominant cropping systems and potential cropping systems in these agroecoregions and their commercial produce, an agro-entrepreneurship with Potential Crop Diversification in Rainfed Areas map (Fig. 13.10) was generated in GIS environment (Background Information, V QRT −2000–2001 to 2004–2005-CRIDA-AICRPDA). Agricultural produce, by and large, is sold in the markets without any postharvest value addition. While the cost of inputs is steadily increasing and that of outputs is either stagnant or reducing, farmers are ending up in loss.

Therefore, postharvest value addition needs an impetus for inducing economic sustainability into farming as an enterprise. Farmers can benefit from better income and more employment opportunities at the village level. Besides, the past decade has shown that dryland horticulture can be a profitable option and has high potential for integrating crop and animal components. This area needs further attention and impetus in the years to come.

13.7   CONCLUSION

Rainfed areas are likely to contribute to nation's food basket, support rural livelihoods and a majority of livestock in India. The biophysical characteristics, particularly climate variability and land degradation, and socioeconomic parameters, particularly adaptive capacity of the resource poor farmers, continue to influence and impact the agricultural land use, thereby production, productivity and profitability. Though, many rural development programs are land based either directly or indirectly linked to rainfed agriculture, the approach has so far been sectoral and piecemeal. Integrated land use planning is a buzzword for achieving the different goals of the various stakeholders, nor single criteria has sustained the land productivities, incomes, ecosystem and finally the livelihoods, the reasons being highly complex situations of risk, diverse socioeconomic settings and subsistence agriculture. However, to achieve the twin goals of sustainable rainfed agriculture and rural development, a integrated land use planning with sound scientific principles is necessary. For this purpose, a new area/target domain approach, that is, Rainfed Agroeconomic zone" is proposed which may address arable, nonarable and common lands with convergence of knowledge, programs/schemes of state and central governments, and activities by relevant stakeholders. While RAEZ is likely to address all the agriculture and related rural development activities in a target domain with judicious mix of natural, capital and human resources, it is suggested to look into some of the available learning experiences, approaches and concepts while operationalizing the RAEZs. A sound Land Use Policy, separately for rainfed regions, at national level, equipped with adequate legislative/judicial powers combined with strong political will, shall promote creation of RAEZs. These RAEZs are contemplated to achieve much desired sustained development and enhance the livelihoods of the farmers in rainfed and drought prone areas and further, pave the way towards achieving "Second Green Revolution" from the rainfed areas.

KEYWORDS

- **Agro-Climatic Regions**
- **Agro-Climatic Zones**
- **Agro-Ecological Regions**
- **Clean Development Mechanism**
- **Desert Development Program**
- **Drought Prone Area Program**
- **Rainfed Agriculture**
- **Rainfed Agro-Economic Zones**
- **Second Green Revolution**
- **Sustainable Agriculture and Rural Development**

REFERENCES

AICRPDA (2003). Annual Reports of 22 Centers (1971–2001). All India Coordinated Research Project for Dryland Agriculture (AICRPDA), Central Research Institute for Dryland Agriculture (CRIDA), Hyderabad, A. P., India, 76357.

Anonymous, (2008). Harmonization of Wastelands/Degraded lands. Datasets of India. Published by National Rainfed Area Authority (NRAA), Ministry of Agriculture, Government o India, NASC Complex, DP Shastri Marg, New Delhi–100012, p. 5.

CRIDA Vision, 2030. CRIDA Vision 2030. Central Research Institute for Dryland Agriculture, Santoshnagar, Hyderabad, India. p. 22.

FAO, (1995). FAO Trainer's Manual, Vol. 1, "Sustainability issues in agricultural and rural development policies.

IPCC (2007). Climate Change (2007). Impacts, Adaptations and Vulnerability. Inter-governmental Panel on Climate Change (IPCC), Working Group II, *www.ipcc.ch*.

NATP-MM-LUP- Rainfed Final Report. (2006). Final Report. NATP-Mission Mode Project on Land Use Planning for Management of Agricultural Resources-Rainfed Agroecosystem. CRIDA, Hyderabad. p. 55.

NATP, (2004). Rainfed Agroecosystem, Production System Research. Completion Report (1999–2004). National Agriculture Technology Project. Central Research Institute for Dryland Agriculture (CRIDA), Hyderabad. p. 202.

Ravindra Chary, G., Venkateswarlu, B., Maruthi Sankar, G. R., Dixit, S., Rao, K. V., Pratibha, G., Osman, M., Kareemulla, K. (2008). Rainfed agro-Economic Zones (RAEZs), A Step towards Sustainable Land Resource Management and Improved Livelihoods. Lead Paper. Proceedings of the National Seminar on Land Resource Management and Livelihood Security, the Indian Society of Soil Survey and Land Use Planning, 10–12 September 2008, Nagpur, India, p. 70.

Ravindra Chary, G., Vittal, K. P. R., Maruthi Sankar, G. R. (2004). A framework of rainfed land use plan for participatory self-sustaining viable bio-diverse mixed farming system. Lead Paper. Proceedings of the National Seminar on Soil Survey for Land Use Planning, 24–25 January 2004, Indian Society of Soil Survey and Land Use Planning, National Bureau of Soil Survey and Land Use Planning (NBSS&LUP), Nagpur, India, p. 2.

Ravindra Chary, G., Vittal, K. P. R., Ramakrishna, Y. S., Sankar, G. R. M., Arunachalam, M., Srijaya, T., Bhanu, U. (2005). Delineation of Soil Conservation Units (SCUs), Soil Quality Units (SQUs) and Land Management Units (LMUs) for land resource appraisal and management in rainfed agro-ecosystem of India: A conceptual approach. Lead Paper. Proceedings of the National Seminar on Land Resources Appraisal and Management for Food Security, Indian Society of Soil Survey and Land Use Planning, National Bureau of Soil Survey and Land Use Planning (NBSS&LUP), Nagpur, India, 10–11 April 2005, p. 212.

Ravindra Chary, G., Vittal, K. P. R., Venkateswarlu, B., Mishra, P. K., Rao, G. G. S. N., Pratibha, G., Rao, K. V., Sharma, K. L., Rajeshwar Rao, G. (2010). Drought Hazards and Mitigation Measures. In: M. K. Jha (editor), Natural and Anthropogenic Disasters: Vulnerability, Preparedness and Mitigation. Capital Publishing Company, New Delhi and Springer, The Netherlands. p. 197–237.

Sehgal, J., Mandal, D. K., Mandal, C., Vadivelu, S. (1992). Agro-ecological regions of India. NBSSLUP Bulletin.No.24,1992. NBSSLUP, Nagpur, India.

Subba Reddy, G., Ramakrishna, Y. S., Ravindra Chary, G., Maruthi Sankar, G. R. (2008). Crop and Contingency for Rainfed Regions of India: A Compendium by AICRPDA. All India Coordinated Research Project for Dryland Agriculture, Central Research Institute for Dryland Agriculture (CRIDA), Indian Council of Agricultural Research, Hyderabad, A. P., India, p. 174.

Velayutham, M., Mandal, D. K., Mandal, Champa, Sehgal, J. (1999). Agro-ecological Subregions of India Fro Planning and Development. NBSS Publ.35, p.372. NBSS&LUP, Nagpur, India.

Vittal, K. P. R., Singh, H. P., Rao, K. V., Sharma, K. L., Victor, U. S., Ravindra Chary, G., Sankar, G. R. M., Samra, J. S., Singh, G. (2003). Guidelines on Drought Coping Plans for Rainfed Production Systems. All India Coordinated Research Project for Dryland Agriculture, Central Research Institute for Dryland Agriculture (CRIDA), Indian Council of Agricultural Research, Hyderabad, A. P., India, p.39.

Vittal, K. P. R., Ravindra Chary, G. (2007). Horizontal and Vertical Diversifications of Rainfed Cropping system in India, in Dryland Ecosystem – Indian Perspective. (Eds) K. P. R. Vittal, R. L. Srivastava, N. L. Joshi, Amal Kar, V. P. Tewari, S. Kathju CAZRI and AFRI, Jodhpur, India. 53–82.

CHAPTER 14

ASSESSMENT AND EVALUATION OF LAND RESOURCES FOR AGRICULTURAL LAND USE PLANNING—A CASE STUDY IN NADIA DISTRICT, WEST BENGAL UNDER IRRIGATED AGRO-ECOSYSTEM

A. K. SAHOO[1], DIPAK SARKAR*, and S. K. SINGH

CONTENTS

[1]National Bureau of Soil Survey and Land Use Planning (ICAR), Block-DK, Sector II, Salt Lake, Kolkata–700091

*National Bureau of Soil Survey and Land Use Planning (ICAR), Amravati Road, Nagpur–440 033

ABSTRACT

The study reports the inventory of natural resources viz. soil, water, climate, land-form, land use, etc., along with socio economic survey in Nadia district (3927 sq.km; AESR 15.1), West Bengal under irrigated agro-ecosystem for development of integrated land use plan of the district. Net sown area in the district is 74% with cropping intensity of 226% and irrigated area is 73% of the net sown area. About 79% of the farmers are marginal (<1 ha land) with an average land holding of 0.55 ha. Four major physiographic units viz. flood plain, meander plain, low lying/marshy area and river valley are identified. 25 soil series are identified during soil resource mapping (1:50,000 scale) and mapped into 43 soil mapping units. Soils are very deep, moderate to poorly drained and coarse loamy to fine in texture. Nine land management units (LMUs) are identified by integrating land units, land use and major production systems. Heavy texture, impeded drainage, moderate flooding, erratic rainfall, poor quality seeds, high cost of agricultural inputs and labor, lack of availability of labor, low prices of products and arsenic contamination in ground water and soil are the major farming constraints. An integrated land use plan is suggested in the district based on soil and land characteristics, crop suitability, farming constraints and SWOT analysis.

14.1   INTRODUCTION

Judicious utilization of land resources with a view to increasing agricultural production to feed the billion plus population on one hand and protecting the environment on the other requires careful land use planning particularly in the background of changing climatic scenario. Alternate land use in conformity with location specific biophysical and socioeconomic environment based upon land evaluation, irrigation potential, detailed soil characteristics and agro-ecological set up is the need of the hour for maintaining the equilibrium between demand and supply as well as mitigating the impact of climate change.

In this backdrop, assessment and evaluation of land resources was carried out in Nadia district, West Bengal under irrigated agro-ecosystem with a view to suggesting an integrated land use plan for optimizing agricultural production in the district.

14.2   STUDY AREA

Nadia district (3927 sq.km), a part of lower Indo-Gangetic Plain in West Bengal and lies between 22052′30″ to 24005′40″ N latitude and 88008′10″ to 88048′15″ E longitude in the alluvial plain of the lower Bhagirathi basin and falls under Agro Ecological Sub Region 15.1 (Bengal basin, hot moist sub humid with deep loamy to clayey alluvium derived soil, medium to high AWC and LGP 210 to 240 days) (Velayutham et al., 1999). The climate of the area is characterized by oppressive hot summer, high humidity, mild winter with annual rainfall ranging from 1424 to 1635 mm. and mean daily minimum and maximum temperature ranges from 80 to 120C (in winter) and 360 to 410C (in summer), respectively.

14.3 METHODOLOGY

Natural resources of the district viz. landform, soil, water, climate, land use, etc., were characterized (Gautam and Murali Krishnan 2005, AISLUS 1970, Soil Survey Staff 2000 and Pal et al., 2009) based on land resource inventory (1:50,000 scale) and various useful thematic maps of the district were generated in GIS.

Primary dataset comprising of soil map, land use map and agro ecological unit map for the district on 1:50,000 scale was prepared for generating LMUs for each block of the district. Land Units (LUs) were delineated by spatial integration of land features, soils, present land use and administrative divisions while LMUs were delineated by integrating land units with prevailing major production systems of the district. Village boundaries of each block was overlaid on the LMU map to generate village-wise statistics of the LMU's.

Village/block-wise socioeconomic data were collected from census records and household data from the farmers covering all the blocks, gram panchayats and collated in conjunction with major soil units of the district for identifying and characterizing the major production systems for developing land use plan of the district. Different components of agriculture and constraints in relation to major farming systems in each LMUs were also identified.

Based on the prevailing agro ecological situation, socioeconomic status and other related factors about agriculture and allied enterprises, SWOT analysis of Nadia district has been made and integrated LUP for different category of farmers in each LMU suggested.

14.4   RESULTS AND DISCUSSION

14.4.1   PHYSIOGRAPHY

Four major physiographic units viz., flood plain, meander plain, low lying/ marshy area and river valley were identified in the district. Meander plains cover the major area (65.2%) followed by flood plain (15.1%), river valley (10.6%) and marshy land (4.0%) (Fig. 14.1). The meander plain is again sub divided into three sub physiographic units viz., upper meander plain, middle meander plain and lower meander plain covering an area of 9.3, 24.5 and 31.4%, respectively.

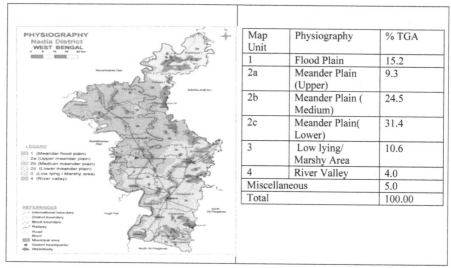

Map Unit	Physiography	% TGA
1	Flood Plain	15.2
2a	Meander Plain (Upper)	9.3
2b	Meander Plain (Medium)	24.5
2c	Meander Plain(Lower)	31.4
3	Low lying/ Marshy Area	10.6
4	River Valley	4.0
Miscellaneous		5.0
Total		100.00

FIGURE 14.1   Physiography of Nadia district.

14.4.2   LAND USE, AGRICULTURE AND IRRIGATION

Data on land utilization for the district revealed that net sown area is 289250 ha (74% of the TGA) while gross cropped area is 656000 ha with cropping intensity of 226%. Irrigated area constituted 73% of the net sown area (Government of West Bengal 2010). The agricultural land is mostly irrigated by river, lift, shallow and deep tube wells. Present land use map of the district is prepared from interpretation of satellite data (Temporal data of IRS P6 LISS III) and five land use units are identified. 79.8% area of the district is under agricultural land (Table 14.1). The major crops grown are paddy (Aman, Aus and Boro), jute, wheat, sesame, sugarcane, mustard, linseed, chilies and wide range of vegetables. The district has high potential for fruit production mainly, mango, banana, papaya, etc., and also has good scope for floriculture.

14.4.3   SOILS

Reconnaissance soil survey of Nadia district was carried out on 1:50,000 scale using the standard procedure. Twenty five soil series are identified in the district and mapped into 43 soil mapping units as soil series association. Soils of the district belong to three soil Orders viz. Inceptisols (66.1%), Alfisols (21.4%) and Entisols (7.5%). Major soil subgroups are Typic Endoaquepts, Aeric Endoaquepts, Vertic

Endoaqualfs, Vertic Endoaquepts, Typic Ustifluvents and Fluventic Haplustepts. The soils are very deep, moderately well to poorly drained and dark grayish brown to gray in color. The pH of the surface soils is slightly acidic to slightly alkaline (5.9–7.6) and in subsurface soils, it ranges from 6.2 to 8.1. Coarse texture in active flood plain areas, imperfect to poor soil drainage, flooding, stagnation of water in low lying areas and arsenic contamination in ground water are the major constraints threatening the production system.

Several themes like surface texture, drainage, particle size, flooding, calcareousness, etc., are generated using the field and analytical data (Table 14.1). Silty clay loam (43.3%) is the most dominant surface texture followed by silty clay (23.5%) and sandy clay loam (10.3%). About 51% area of the district suffers from moderate to severe flooding hazard. Imperfect to poorly drained soils are found in 82.3% area of the district and moderate to strong calcareous soils are observed in 34.1% area of the district.

TABLE 14.1 Status of Present Land Use, Flooding, Surface Soil Texture, Soil Particle Size, Soil Drainage and Calcareousness

Themes	Class	% TGA
Present Land Use	Agricultural land	79.8
	Forest	0.1
	River	2.1
	Wet land	4.0
	Habitation	13.6
Flooding	No flooding (F0)	4.1
	Slight/occasional flooding (F1)	40.3
	Moderate flooding (F2)	39.9
	Severe flooding (F3)	10.7
Surface Soil Texture	Clay	6.1
	Silty clay	23.5
	Silty clay loam	43.3
	Sandy clay loam	10.3
	Loam	2.9
	Silt loam	2.7
	Sandy loam	6.2

TABLE 14.1   *(Continued)*

Themes	Class	% TGA
Soil Particle Size	Very fine	24.5
	Fine	20.7
	Clayey over fine silty	20.5
	Fine loamy	5.1
	Fine silty	6.7
	Coarse loamy	5.6
	Coarse silty	5.1
	Clayey over loamy	3.1
	Fine loamy over sandy	3.5
Internal Soil Drainage	Poor	50.5
	Imperfect	31.8
	Mod. Well	12.5
	Well	0.2
Soil Calcareousness	Slightly calcareous	30.0
	Moderately calcareous	24.3
	Strongly calcareous	9.8
	Nil	30.9

14.4.4   LAND MANAGEMENT UNITS

Twenty land units (LUs) were delineated by spatial integration of land features, soils, present land use and administrative divisions. Subsequently, nine Land Management Units (LMUs) were identified (Fig. 14.2) by integrating 20 Land Units with prevailing four major production systems viz. cropping system (jute-rice/rice-rice/mustard/sesame/wheat/gram/black gram/lentil/vegetables/flowers, sugarcane, banana, etc.), live stock (dairy, poultry, duckery, goatery, piggery), aquaculture (fishery) and homestead (vegetables, fruits, flowers, betel-vine, etc.). The salient features of the LMUs are presented in Table 14.2.

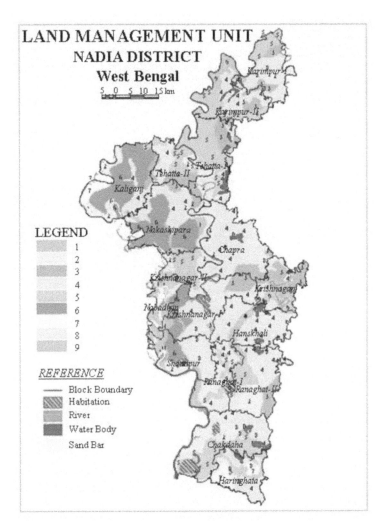

FIGURE 14.2 Land Management Unit of the District.

TABLE 14.2 Salient Features of LMUs of Nadia District, West Bengal

LMUs	Area (ha)	Landform	Soil Characteristics
1	13,476	Nearly level to very gently sloping flood plain	Moderate to imperfectly drained calcareous fine silty soils
2	52,463	Nearly level flood plain	Imperfectly drained calcareous coarse silty soils

TABLE 14.2   *(Continued)*

LMUs	Area (ha)	Landform	Soil Characteristics
3	33,939	Nearly level to very gently sloping meander plain	Moderately drained calcareous fine soils
4	113,478	Very gently sloping meander plain	Poor to imperfectly drained fine soil
5	62,238	Very gently sloping meander plain	Poorly drained calcareous fine loamy soils
6	46,844	Nearly level to very gently sloping meander plain	Poorly drained fine silty soils
7	6356	Very gently sloping meander plain	Poorly drained calcareous fine silty soils
8	23,986	Nearly level to very gently sloping marshy land	Poorly drained fine soils with severe flooding
9	17,482	Very gently sloping river valley	Poorly drained fine loamy soils

Soils of LMU 1 (13,476 ha) and LMU 3 (33939 ha) are moderately to imperfectly drained while rest of the soils suffered from poor drainage.

14.4.5. SOIL-SITE SUITABILITY FOR CROPS

Soil site suitability evaluation (Sys et al., 1993) for various crops in relation to cropping pattern of the district under each LMU revealed that soils of 66%, 94% and 86% area of the district are moderately suitable (S2) to highly suitable (S1) for rice, jute and mustard/rapeseed, respectively, while about 94% area are moderately (S2) to marginally (S3) suitable with respect to wheat, sesame and sugarcane cultivation. About 89% area is moderately (S2) to marginally (S3) suitable for groundnut cultivation (Table 14.3).

TABLE 14.3   Soil Site Suitability for Different Crops in Nadia district, West Bengal (% TGA)

Suitability class	Rice	Jute	Mustard/ Rapeseed	Wheat	Sesame	Sugarcane	Groundnut
S1	22.0	15.4	9.5	0	0	0	5.1
S2	43.6	78.9	78.7	19.3	41.8	19.8	29.7
S3	28.7	0	6.1	75.0	52.5	74.5	59.5

14.5 SOCIO ECONOMIC STATUS AND MAJOR FARMING CONSTRAINTS

Comprehensive socioeconomic appraisal evaluated for each LMUs based on farm/household data pertaining to 17 blocks of the district revealed about 79% of the farmers as marginal (<1 ha) with 18% falling under small category (1–2 ha) with an average holding of 0.55 ha and 1.69 ha, respectively (Government of West Bengal, 2010). Poor quality seeds, high cost of labor and electricity and erratic rainfall are the major farming constraints identified in the district. Heavy texture with impeded drainage, moderate flood hazard resulting in loss of fertile surface soils, high cost of agricultural inputs, low prices of product and lack of availability of labor are also some of the constraints identified (Table 14.4). Arsenic contamination in ground water and soil is also reported.

TABLE 14.4 Farming Constraints Affecting the Crop Yield in Nadia District, West Bengal

LMU (No. of Farm families)	Percent of Farmers (Ranking of Problem)							
	Soil	Rainfall	Seeds	Fertilizer	Labor	Electricity	Cost of inputs	Prices of product
LMU 1 (20)	60 (8)	80 (4)	100 (3)	70 (7)	100 (1)	100 (2)	70 (6)	80 (5)
LMU 2 (42)	88 (7)	100 (4)	100 (3)	82 (8)	100 (2)	100 (1)	90 (5)	88 (6)
LMU 3 (25)	76 (7)	100 (2)	92 (4)	76 (8)	88 (5)	100 (1)	100 (3)	80 (6)
LMU 4 (47)	77 (5)	100 (2)	100 (1)	74 (6)	91 (4)	93 (3)	72 (7)	67 (8)
LMU 5 (34)	33 (8)	100 (4)	100 (3)	59 (7)	100 (1)	100 (2)	94 (6)	100 (5)
LMU 6 (29)	72 (8)	93 (4)	100 (1)	72 (7)	96 (2)	93 (3)	76 (6)	86 (5)
LMU 7 (18)	67 (8)	100 (4)	100 (3)	78 (7)	100 (1)	100 (2)	100 (5)	88 (6)
LMU 8 (30)	80 (8)	90 (4)	100 (2)	80 (7)	100 (1)	90 (3)	87 (5)	80 (6)
LMU 9 (22)	45 (8)	95 (4)	95 (5)	73 (7)	100 (2)	100 (1)	100 (3)	82 (6)

14.6   SWOT ANALYSIS

Based on prevailing agro-ecological situation, socioeconomic status and other related factors about agriculture and allied enterprises, SWOT analysis of Nadia district has been made. The strengths, weakness, opportunities and threats identified are presented in Table 14.5.

TABLE 14.5   Strengths, Weakness, Opportunities and Threats, Nadia district, West Bengal

Strengths	Weakness	Opportunities	Threats
• Nearly level to very gently sloping alluvial plain land • Fertile soils • Well irrigation facility (73 percent of the net sown area) Multiple cropping systems • Higher educational and literacy level (more than 60 percent) • Well-developed co-operative sector • Well-established communication system • Vast multidisciplinary research infrastructural support offered by BCKV, Mohanpur, Nadia, West Bengal • High cosmopolite nature of the farming community	• Uneven distribution of rainfall • Unassured irrigation due to frequent power cuts • Fragmented land holding • High cost of labor, electricity and agricultural inputs • Poor quality of seeds, also not timely available • Increasing rate of degradation in natural resources • Low aspiration towards adoption of technologies	• Ample scope for utilization of raw material to promote agro based industries • Diversification in agriculture and other enterprises • Market driven extension to improve yield and quality of various commodities based on customer need • Use of advanced information technology • Communication for environment friendly technologies and concepts • Enhance application of user-friendly biotechnological options • Awareness for technology development and dissemination	• Small land holding sometimes affects the adoption of technologies • Natural calamities, fluctuating price policies, hike in inputs costs influence in adoption of technologies • Arsenic contamination in ground water and soils.

14.7   SUGGESTED LAND USE PLAN

Based upon soil and land characteristics, climate and soil-site suitability for different crops, farming constraints affecting the crop yield, need/benefit of the farmers and SWOT analysis, an integrated land use plan for different category of farmers for each LMUs of the district is suggested towards optimizing agricultural production as well as environmental protection (Table 14.6)

TABLE 14.6 Suggested Land Use Plan in Different LMUs of Nadia District, West Bengal

Constraints	Alternate options for Marginal Farmers	Alternate options for Small Farmers	Alternate options for Medium Farmers
LMU-1			
1. Moderate to severe flood hazards and consequent loss of fertile surface soils. 2. Lack of availability of labor, fertilizers and good quality seeds. 3. Late onset and intermittent dry spells. 4. Lack of fodder availability for livestock. 5. Arsenic contaminated groundwater.	Jute – kharif paddy –vegetables/ groundnut/flowers along with integrated fish paddy farming during kharif season in low lying area with backyard livestock (Dairy/ Poultry/ Duckery + Goatery).	Jute – kharif paddy-vegetables/mustard/rapeseed along with integrated fish paddy farming during kharif season with backyard livestock (Dairy/Poultry/Duckery + Goatery)	*Integrated Farming system with Rain-fed paddy and backyard Poultry/Duckery/Goatery/Dairy with various cropping sequences. *Jute – kharif paddy –mustard/wheat/sesame/lentil/gram *Sugarcane, Banana
	Homestead: Mango/guava/papaya/pine apple plantation with vegetable in both kharif and rabi seasons andsite specific integrated nutrient, pest and crop management.		
LMU-2			
1. Severe flood hazards and consequent loss of fertile surface soils. 2. Prolonged inundation in depressed lands restricts cultivation. 3. Lack of availability of labor and seeds. 4. Low productivity of rain-fed paddy in coarse textured soils. 5. Infectious disease in livestock. 6. High cost of inputs.	Jute – kharif paddy –potato/ vegetables/ marigold along with integrated fish paddy farming during kharif season in low lying area with backyard livestock (Dairy/Poultry/ Duckery + Goatery)	Jute – kharif paddy-vegetables/ potato/ mustard rapeseed/ along with integrated fish paddy farming during kharif season with backyard livestock (Dairy/ Poultry/ Duckery + Goatery)	*Jute – kharif paddy – vegetables/ wheat/ mustard/ sesame/lentil along with integrated fish paddy farming with livestock. *Sugarcane, Banana
	Homestead: Mango/guava/papaya/ banana plantation, green gram, black gram,betel-vine, potato and vegetable in both kharif and rabi seasons with site specific INM and IPM.		

LMU -3			
1. Fine heavy texture of sub-soil layers resulting compaction in dry season limit crop growth. 2. Lack of availability of labor, fertilizers and good quality seeds.	Jute – kharif paddy –chili/gram/ mustard/marigold along with backyard livestock (Dairy/ Poultry/ Duckery + Goatery) and fishery.	Jute/paddy – kharif paddy- chili/gram/ mustard/marigold along with integrated fish paddy farming in low lying area during kharif season and backyard livestock.	*Jute/paddy – kharif paddy – mustard/ rapeseed/ sesame/ lentil/ gram/ sunflower along with integrated fish paddy farming with livestock. *Intercropping with banana plantation.
3. Poor livestock production due to lack of feed and fodder. 4. Irregular rainfall. 5. Less prices of products.	Homestead: Papaya, ginger, chili and other vegetables in both kharif and rabi seasons with site specific integrated nutrient and pest management.		
LMU-4			
1. Heavy texture with impeded drainage. 2. Late onset and intermittent dry spell. 3. Attack of pest during rain-fed paddy. 4. Lack of availability of seeds, fertilizers and labors.	Jute – kharif paddy –mustard/ chili/ marigold along with backyard livestock and fishery.	Jute/paddy – kharif paddy – mustard rapeseed/ sunflower along with integrated fish paddy farming during kharif season with backyard livestock (Dairy/ Poultry/ Duckery + Goatery).	*Jute/paddy – kharif paddy – mustard/ sesame/lentil/ sunflower along with integrated fish paddy farming with livestock. *Intercropping with banana plantation.
5. Lack of irrigation water during rabi season. 6. Arsenic contaminated groundwater.	Homestead: Mango/guava/banana plantation with vegetable in both kharif and rabi seasons andsite specific nutrient and pest management.		
LMU-5			

1. Late onset and intermittent dry spell. 2. Lack of irrigation water for rabi crops. 3. Lack of availability of labor and good quality seeds. 4. Less prices of products. 5. Arsenic contaminated groundwater.	Jute – kharif paddy –potato/ vegetables/ marigold/ tuberose along with integrated fish paddy farming during kharif season in low-lying area with backyard livestock.	Jute – kharif paddy-vegetables/ chili/ mustard/ rapeseed along with integrated fish paddy farming during kharif season with backyard livestock.	*Jute – kharif paddy- vegetables/ chili/ mustard/ rapeseed/ wheat/ gram/ sesame with backyard (Poultry/ Duckery/ Goatery/ Dairy) and fishery. *Sugarcane and Banana plantation.
	Homestead: Guava/papaya/lichi plantation, green gram, black gram, betel –vine and vegetable in both kharif and rabi seasons with site specific integrated nutrient, pest and crop management.		
LMU-6			
1. Moderate flood hazards and consequent loss of fertile surface soils. 2. Impeded soil drainage. 3. Poor livestock production due to lack of feed and fodder. 4. Lack of irrigation water for rabi crops 5. Lack of availability of labor and good quality seeds.	Jute – kharif paddy –vegetables/ potato/ chili/ flowers along with integrated fish paddy farming during kharif season in low lying area with backyard livestock (Dairy/ Poultry/ Duckery + Goatery).	Jute – kharif paddy-vegetables/ mustard/ sesame/ rapeseed along with integrated fish paddy farming during kharif season with backyard livestock (Dairy/ Poultry/ Duckery + Goatery)	*Integrated jute – paddy mustard/ rapeseed/ sesame/ sunflower/ lentil/ gram and backyard livestock and fishery. *Sugarcane, Banana
	Homestead: Mango/papaya/ banana plantation with potato, vegetable in both kharif and rabi seasons and site specific integrated nutrient and pest management.		
LMU-7			
1. Moderate flood hazards. 2. Impeded soil drainage. 3. High labor wages. 4. Lack of fodder availability of livestock. 5. Low productivity of rain-fed paddy. 6. High cost of inputs.	Jute – kharif paddy –potato/ vegetables/ marigold along with integrated fish paddy farming during kharif season in low lying area with backyard livestock (Poultry/ Duckery + Goatery)		
	Homestead: Mango/guava/papaya/lichi plantation, betelevine, ginger, potato and vegetables in both kharif and rabi seasons with site specific integrated nutrient, pest and crop management.		
LMU-8			

1. Prolonged inundation restricts cultivation.	*Integrated fish – paddy/ jute farming system during kharif season with backyard livestock.
2. Heavy texture with impeded drainage.	*During rabi season, vegetables/mustard/ rapeseed/sesame/ sunflower.
3. Poor livestock production due to infectious diseases.	
4. Attack of pest during rain-fed paddy.	Homestead: Jackfruit/ banana plantation and vegetables in both kharif and rabi season-swithsite specific integrated nutrient and pest management.
5. Lack of availability of labors and fertilizers.	
6. Wetlands are badly affected by wild shrubs.	
LMU-9	
1. Impeded soil drainage condition.	Jute – kharif paddy –potato/ vegetables/ marigold/ tuberose. Integrated fish paddy farming during kharif season in low lying area with backyard livestock.
2. Severe flood hazards and loss of surface fertile soils.	
3. Prolonged inundation in depressed lands restricts cultivation.	
4. Lack of availability of labor and good quality seeds.	Homestead: Papaya plantation with black grams, green grams and vegetables and site specific integrated nutrient and pest management.
5. Arsenic contaminated groundwater.	
6. Lack of availability of fodder.	

14.8 CONCLUSION

Soils of Nadia district, West Bengal developed on nearly level to very gently sloping land of Indo-Gangetic plain (IGP) are found favorable for growing wide variety of crops. However, fractional land holdings, erratic rainfall, poor quality seeds, high cost of agricultural inputs and labor, lack of availability of labor, low prices of products and arsenic contamination in ground water and soil are the major constraints hampering the cultivation of different crops. An integrated land use plan for different categories of farmers under different land management units is suggested towards optimizing agricultural production as well as environmental protection.

KEYWORDS

- **Indo-Gangetic Plain**
- **Land Management Unit**
- **Land Units**
- **Nine Land Management Units**
- **SWOT analysis**

REFERENCES

AISLUS (1970). Soil Survey Manual, IARI, New Delhi.

Gautam, N. C., Murali Krishnan, I. V. (2005).Natural Resource Classification System, Centre for Land Use Management, Hyderabad, India.

Government of West Bengal (2010). District Statistical Hand Book (2008). Nadia, Bureau of Applied Economics and Statistics, Government of West Bengal. 96p.

Pal, D. K., Tarafdar, J. C., Sahoo, A. K. (2009). Analysis of Soils for Soil Survey and Mapping. In Soil Survey Manual (T. Bhattacharyya, et al. Ed.), NBSS & LUP Publication No.146. India, 83–145.

Soil Survey Staff (2000). Soil Survey Manual, United States Department of Agriculture, Hand Book No.18, Reprint in India by Scientific Publishers (India), Jodhpur.

Sys, C., Van Ranst, E., Debaveye, J. (1993). Land Evaluation. Part 3: Crop requirements, Agricultural Publications 7,3. General Administration of Development Cooperation of Belgium, Brussels. 199p.

Velayutham, M. Mandal, D. K., Mandol, Champa, Segal, J. (1999). Agro-ecological Sub regions of India for Planning and Development. NBSS Publ.35, NBSS & LUP, Nagpur, India 372 p.

CHAPTER 15

MICRO-LEVEL PLANNING FOR SUSTAINABLE RESOURCE MANAGEMENT IN AGRICULTURE—A KERALA PERSPECTIVE

A. PREMA[1] and JIJU P. ALEX

CONTENTS

[1]Associate Professor (Agricultural Economics), College of Horticulture, Kerala Agricultural University.

ABSTRACT

Diversity in the microenvironment imposes severe limitations to introduce a plan for the farming systems. A systematic assessment of physical, social and economic factors is necessary to encourage and assist land users in selecting sustainable options that increase their productivity and meet the needs of society. Information at the microlevel is required to design sustainable land use models and technologies. This chapter outlines an attempt taken up through stakeholder participatory approach on an agro-ecological unit basis. It deliberates on approaches to identify the constraints and potential in farming and suggest broad agricultural development plans, in Kerala. A SWOT analysis of the agricultural production systems indicated wide variations in the physical, sociopolitical, human and financial endowments. The projects on agriculture implemented by the local bodies were often found to lack comprehensiveness and integrated nature. Specific agro-ecological unit based interventions with respect to crop, livestock production and other agro-related enterprises and spatial integration of crop is warranted. Integration of other line departments and major programs like NREGS under the aegis of Krishibhavans with the support of local bodies is suggested as the strategy for the development of agriculture in Palakkad.

15.1   INTRODUCTION

"In order to further revive agricultural growth in the states, agro-ecological planning/agro-climatic planning could be adopted in a decentralized planning framework to optimally use the zonal potential/resources in crop and livestock production" (Planning Commission, GOI, 2011).

Diversity in the microenvironments imposes severe limitations to introduce a plan for the farming system as a whole. Over the years, it has been realized that, there is a substantial gap between, generated technologies and sensitiveness of agricultural technologies to the agro-ecological situations, social factors, financial factors and preferences of farm families. Development of appropriate packages that are adaptable to different agro-climatic zones was a strategic step towards increasing accessibility to new adoptable technologies in resource poor situations. Information at the microlevel is required to design sustainable land-use models and technologies that enable optimal utilization of the local resources—both bio-physical and economic. A systematic assessment of land and water potential, alternatives for land use, and economic and social conditions is necessary for selection and adoption of the best land-use options. Hence, it is imperative to assess the development potential of the dominant sector of the economy, that is, agriculture. Several attempts made to study the agrarian structure and economic development has failed to identify the prime reasons behind the dismal performance of the farm sector especially at the micro level. Attempts to assess the suitability of microclimatic and region specific socioeconomic parameters in farming are often staggered and non comprehensive. In fact, there have been not many concerted efforts to revive farming or to develop new plans in the light of the results and conclusions of the microlevel studies. Moreover, the substantial gap existing between generated technologies and sensitiveness of agricultural technologies to the agro-ecological situations, social factors, economic factors and preferences of farm families have often been overlooked.

Non-availability of accurate and dependable database below the taluk level on agriculture and allied sector has been pin-pointed as one of the major constraints impeding resource-based planning by the Planning Commission, GOI (2011). They have rightly pointed out the need for initiating comprehensive exercise at state level to delineate the state into various agro-ecological/agro- climate zones and the districts into further subunits. In each agro-ecological unit, resource based plan need to be developed including issues like yield gap also. The agro-climatic Regional Planning exercise of the Planning Commission during 8th and 9th plan period was a serious attempt to plan for smaller homogenous regions, keeping in view the natural resources and capabilities towards achieving development.

The Agenda 21 Document of the Rio Conference (1992) upholds the basic right of people to be involved in decision-making exercises, which directly affect them. This chapter outlines an attempt taken up through stakeholder participatory approach to identify the constraints and potential of agricultural enterprises and to suggest broad agricultural development plans, in Kerala on an agro-ecological unit level in one of the agriculturally potential district of Kerala—the Palakkad. The problems and constraints to farming were assessed along with the strength and opportunities to develop farming into a sustainable activity on the basis of resource endowments in the respective agro-ecological units. Enterprise mixes were suggested for the agro-ecological units based on stakeholder preferences and suitability. The study complemented the delineation of Palakkad district into different agro-ecological units based on soil and climatic considerations by NBSS&LUP, Bangalore as part of RSVY (Fig. 15.1).

FIGURE 15.1   Map of Palakkad district.

15.2   PALAKKAD—THE RICE GRANARY OF KERALA

Palakkad district is located in the east central portion of Kerala state and lies between $10°\ 19'$ and $11°\ 14'$ North latitudes and $76°\ 1'$ and $76°\ 54'$ East longitudes. The district is generally a plain surrounded by Sahyadri and Nilgiri ranges. These mountain ranges have got a gap of about 30 km, known as the Palakkad gap, which has considerable influence over the climatic condition of the region. The district has humid tropical climate with temperature ranging from 19°C to 42°C.

The South-west monsoon begins in June and lasts till September while, Northeast monsoon shower is received during October to January. About 75% of the total rainfall is received during South-west monsoon. Palakkad district occupies, 11.5% of the total geographical area of the State (4457.84 sq.km). A large portion of the district is covered by reserve forests. According to the Census 2001, the total population of the district is 26,17,072 and the growth rate of population is 9.86, which is higher than that of the State (9.42). Density of population per sq. km in this district is 584 and the sex ratio is 1068. The literacy rate is 84%. The per capita income of the district is Rs. 27,649, which is below the state average (Rs. 29,618). Agriculture and allied sectors, contribute Rs. 1,02,662 lakhs (at current prices) to the Gross Domestic product of the district. This accounts for 8.1% of State GDP. Out of the total 8,45,181 workers in the district, 3,48,299 (41%) are agricultural laborers. Palakkad district is divided into two Revenue Divisions- Palakkad and Ottappallam and five taluks viz., Palakkad, Chittur, Alathur, Ottappallam and Mannarkkad and 156 villages. The district consists of four municipalities, 13 block panchayats and 90-one grama panchayats.

Palakkad is considered to be the "Rice Granary" of the state. The district accounts for about 38% of paddy area producing about 34% of the total rice production in the state (Economic Review, 2010). The land utilization pattern of the district (Table 15.1) indicates that the district has 28,356 hectares of wastelands under various categories, which is nearly 6% of the total geographical area. The district has diversified units of topography, physiography, climate and soil and land-use pattern. The major crops cultivated in the district are paddy, pulses, coconut, sugarcane, mango, jack, vegetables, banana and plantains. Cropping of cotton, groundnut and ragi in the state is exclusively confined to this district.

TABLE 15.1 Land Utilization Pattern in Palakkad District, Kerala 2009–2010

Particulars	Area (Lakh ha)	% share to corresponding state area
Total Geographical area	4.47	11.5
Forest	1.36	12.6
Land put to non agricultural use	0.43	11.43
Barren and uncultivable land	0.034	13.38
Pastural/Grazing land	0	-
Land under miscellaneous tree crops	0.015	34.09
Cultivable waste	0.28	28.57
Fallow other than current fallow	0.10	22.22
Current fallow	0.11	14.29
Water Logged Area and Marshy land	0	-
Still water	0.15	14.70
Social forestry	0.0003	1.0
Net area sown	22.07	10.63
Total cropped area	3.14	11.76

Source: Department of Economics and Statistics, Kerala.

The area, production and productivity of major crops in the district are presented in the Table 15.2. Paddy is the most important crop occupying the largest area under cultivation accounting for 43% of the total area under cultivation and 44% of paddy produced in the state during 2009–2010. More than half the area and production of pulses in the state is in Palakkad. Yet, its productivity is very low. Plantation and horticulture crops account for about 45% of the total cropped area in the district. There had been a steady increase in the area under coconut cultivation over the years. However, during the last three to four years, a decreasing trend in area expansion under coconut was observed mainly due to the fall in the prices of coconut. The area under mango plantation in the district is estimated at 7701 hectares (constituting nearly 12% of state share) with a production of 73,313 tons, which is higher by nearly 60% over the state average. Most of the mango orchards are concentrated in a particular village – Muthalamada, which is now popular as 'mango city.'

Paddy, coconut and rubber together contribute to nearly 80% of the agricultural income of the district (Table 15.2). It is paradoxical to note that the income contribution from paddy (Rs. 38,920 lakhs) comes from a total area of 1,00,522 hectare and that of rubber (Rs. 51,349 lakhs) from an area of 47930 hectare accounting to 43.32 and 6.4% of the aggregate income of the state from the respective crops. The income contribution from paddy is Rs. 38,726 per hectare whereas as that from rubber is Rs. 1,07,133/- per hectare. Coconut contributes to Rs. 42,297 per hectare to the agricultural income of the district (Source: Agricultural Statistics 2011–2012).

TABLE 15.2   Area, Production and Productivity of Major Crops in Palakkad District, Kerala

S. No.	Crop	Area (ha)	Production (tones)	Productivity (kg/ha)
1	Paddy	100,522 (42.96)	266,231 (44.50)	2648
2	Coconut	57,186 (7.34)	417 (million nuts) (7.36)	7292 (nuts per ha)
3	Pulses	2986 (67.11)	2244 (66.19)	752
4	Tapioca	2843 (3.58)	90,428 (4.1)	31,807
5	Rubber	35,559 (6.77)	47,930 (6.42)	1348
6	Mango	7701 (12.08)	73,313 (19.64)	9520
7	Banana	10,593 (20.66)	80,068 (19.70)	7559
8	Cashew	3002 (6.13)	1047 (2.92)	349

Figures in parentheses show percent share to corresponding state total.
Source: Department of Economics and Statistics, Kerala

The district also has a sizeable livestock wealth and accounts to nearly 15% of the bovine population of the state. The production of eggs in the district is 1070 Lakh against the annual requirement of 8700 lakh (Source: District Credit plan— 2006–2007). The district ranks first with respect to fresh water area brought under fish culture with assistance from Fish Farms Development Agency (FFDA).

Irrigation projects play a major role in retaining the agricultural importance of the district. There are three major and five medium irrigation projects. Nearly 33,512 hectares of land have been brought under major irrigation schemes. The main source of irrigation is canal and the district accounts for about 46% of the total area irrigated by canals in the State.

15.3   PARTICIPATORY STAKEHOLDER ANALYSIS

People's participation is very important for the success of a development project. Participatory tools are frequently made use off, for assessing various development issues such as local resources, watershed issues, environmental problems, local needs and gender issues in local communities in a relatively short period. This study uses stakeholder participatory approach originally propounded and propagated by Chambers (1991) with appropriate modification to identify the constraints and potential of the selected agro-ecological units of Palakkad district. Matrix ranking, problem analysis chart and SWOT analysis were resorted to explore the prevailing situation and to suggest broad development plans for the agricultural production systems in the selected agro-ecological units in the district, viz., Unit VII, Unit VIII, Unit IX, and Unit X.

Taking in to consideration the geographical condition and logistic convenience of farmers, 3–4 PRA sessions were conducted in each zone, pooling panchayats. Key informant farmers (KIF) representing each of the enterprises, elected representatives of panchayats, officials of the departments of Agriculture, Animal husbandry, Dairy development and Irrigation attended the PRA sessions. Semi-structured schedules and guidelines were prepared for facilitating the sessions. The data collected were subjected to triangulation at a workshop of stakeholders organized at district head quarters.

15.4   STAKEHOLDER PREFERENCE FOR ENTERPRISE—MIX

With rising population and declining land-man ratio, agriculture may not be able to provide adequate income and employment to households in the long run. Integration of farm enterprises suitable to different agro-ecological areas and their resource endowments would provide better livelihoods to the farmers. The enterprise suitability of the selected agro-ecological units of Palakkad was studied in terms of profitability, marketability and resource availability through enterprise preference matrix prepared by the villagers (Table 15.3). The study clearly indicated that, the preference for enterprises varied across different units within the district. Although Palakkad is referred to as 'rice granary,' paddy has been ranked low by the farmers in the enterprise preference matrix. The percent difference in profit from paddy and the alternative crops varied from Rs. 108 to Rs. 1900 per hectare (Rajalekshmi, 2006).

A comparative analysis of the strengths, weaknesses, opportunities and threats (SWOT) of the selected units indicated that the units varied from each other in terms of inherent strengths and weaknesses. Favorable climate and fertile soil suited to diverse crops, presence of irrigation and water sources are the major strengths of all units. Large cattle population and availability of water bodies to sustain an integrated farming system are added advantages. The functional padasekara samithis and Kudumbasree units gives impetus to agricultural developmental activities in all the four units. The National Rural Employment Guarantee Scheme (NREGS) being operational in the district since 2006 could be effectively harnessed to solve the labor scarcity problems in agriculture (Table 15.4).

Apropos weaknesses, it could be seen that, even though sufficient water and irrigation sources are present, lack of proper maintenance of irrigation canals, silting of dams and unscientific scheduling of irrigation water have led to the inefficiency of resource use. Capital inadequacy and labor scarcity coupled with nonavailability of agro machineries suitable to small holdings have increased the cost of cultivation. Even though the units have substantial cattle population, livestock management on scientific basis is not a common knowledge among farmers. The farmers in all the units pointed out the insufficiency of financial assistance from the Government for farming (Table 15.5). The opportunities of these units for enhancing production, productivity and avail gainful employment are considerably high, provided adequate and specific interventions are made in each unit (Table 15.6). The prospects of harnessing the opportunity through the established network of Self-help groups of Kudumbasree and the like are yet to be realized. Specific projects that tap the opportunities have to be formulated covering all the prospective agro- ecological units.

The major threats identified through the study, suggest location specific and policy level interventions to create conducive environment for growth and development (Table 15.7). Even though the Kerala Land Utilization Act is in force, the loopholes in the Act and the lack of emphasis in its enforcement, still lead to conversion of paddy fields to other crops and nonagricultural purposes. The apathy and dying interest of people, particularly youth in farming is yet another serious threat to agriculture. Concerted efforts to retain people in farming through innovative approaches and assurance of remunerative and stable price are needed. Of late, the impact of climate change on farming and farm income has been more pronounced. Suitable cropping sequence and crop calendar have to be adopted by farmers to combat climate change. Expedition of attractive crop insurance schemes by the Government is the need of the hour, lest the farmers would be at peril.

TABLE 15.3   Crop/Enterprise Suitability Matrix

S. No.	Crop/Enterprise	Unit VII	Unit VIII	Unit IX	Unit X
1	Paddy	V(23.5)	VII(18)	VIII(12)	III(25)
2	Rubber	-	I (26.5)	I(28)	I(28)
3	Dairying	I(28.5)	I(26.5)	VI(17)	VII(21.5)
4	Vegetables	IV(24)	III(24.5)	-	VIII(20)
5	Poultry	VI(23)	III(24.5)	-	-
6	Tapioca	V(23.5)	II (26)	IV(21.5)	VI(22)
7	Banana	-	III(23)	IV(21.5)	II(26)
8	Goat rearing	-	IV(22)	II(24)	IV(23.5)
9	Agro processing	-	-	III(22.5)	VII(21.5)
10	Apiculture	-	-	III(22.5)	IV(23.5)
11	Sericulture	II(26)	-	-	-
12	Toddy tapping	I(28.5)	-	-	-
13	Mango cultivation	III(24.5)	-	-	-

TABLE 15.4   Comparative Analysis of Strengths of the Selected Agro Ecological Units

Strengths	Unit VII	Unit VIII	Unit IX	Unit X
Fertile soil and congenial climate	✓	✓	✓	✓
Presence of irrigation and water sources	✓	✓	✓	✓
Availability of ponds, water bodies etc.	✓	-	-	✓
Large cattle population		-	✓	-
Traditional rice mills	✓	✓	-	-
Majority of farmers are marginal or small farmers	✓	-	-	-
Existence of functional padasekharams	✓		✓	-
Access to technology		-	-	✓
Marketing network of VFPCK	-	✓	-	-
Plantation crops dominated farming	-	-	-	✓
Under utilized homesteads	-	-	✓	✓
Functional Kudumbasree/SHG	-	✓	✓	✓
Presence of Functional Farmers Service Cooperative Banks	-	✓	-	-
Paddy farming as a major livelihood activity	✓		-	✓
NREGS in operation	✓	✓	✓	✓

TABLE 15.5   Comparative Analysis of Weaknesses of the Selected Agro Ecological Units

S. No.	Weaknesses	VII	VIII	IX	X
1	Lack of proper maintenance of irrigation canals/ponds	✓	✓	✓	✓
2	Unscientific irrigation scheduling	✓	✓		
3	Scarcity of water	✓			✓
4	Insufficiency of agromachinaries	✓		✓	✓
5	Delay in supply of paddy seeds	✓	✓		✓
6	Insufficient veterinary care	✓		✓	
7	Fodder availability is less		✓		
8	Insufficient milk collection centers	✓		✓	
9	Non availability of timely labor	✓	✓	✓	✓
10	Lack of awareness on scientific/cattle management	✓		✓	
12	Milk societies unable to collect surplus production	✓			
13	Non functioning mills/factories	✓	✓		
14	Low adoption of recommended crop/livestock management practices	✓		✓	
15	Lack of systematic pest and disease control				✓
16	Dependence on informal credit	✓			
17	Lack of storage and marketing facilities		✓	✓	✓
18	Unstable price of agricultural produces		✓		✓
19	No marketing support for kudumbasree			✓	
20	Poor performance of RSGP		✓	✓	
21	Insufficient technical support in field			✓	✓
22	Large variation in land holding size				✓
23	Insufficient financial assistance from Government	✓	✓	✓	✓

TABLE 15.6   Comparative Analysis of Opportunities of the Selected Agro Ecological Units

S. No.	Threat	VII	VIII	IX	X
1	Presence of several agro processing units and traditional mills	✓	✓		
2	Under utilized coconut gardens and homesteads	✓	✓	✓	✓
3	Fallow wetlands for second crop	✓		✓	✓
4	Summer rice fallows		✓	✓	✓
5	Fish farming in ponds	✓		✓	✓

TABLE 15.6 *(Continued)*

S. No.	Threat	VII	VIII	IX	X
6	Prospects of dairying		✓	✓	✓
7	Waste lands for fodder cultivation		✓		
8	Prospects of Rabbit/Goat/Poultry	✓	✓	✓	✓
9	Prospects of micro enterprises through groups	✓	✓	✓	✓
10	Agroservice centers of youth and custom hire service	✓			
11	Functional group farming committees	✓	✓	✓	
12	Integration with MGNREGS	✓	✓	✓	✓

TABLE 15.7 Comparative Analysis of Threats for Development in the Selected Units

S. No.	Threat	VII	VIII	IX	X
1	Uncertainity with regard to release of water from Parambikulam–Aliyar project	✓	✓		
2	Inflow of inferior quality planting materials and agricultural inputs from neighboring states	✓			
3	Short supply of Potash fertilizers	✓			
4	Natural calamities, pest and diseases, climate change	✓	✓	✓	✓
5	Damage by wild animals and birds	✓	✓	✓	✓
6	Undulating topography rendering farm mechanization difficult				✓
7	Upper limit of land ceiling for availing benefits	✓	✓		
8	Loopholes in KLU Act and its non effective enforcement				✓
9	Flash bandhs and hartals	✓	✓	✓	✓
10	Indebtedness of farmers				✓
11	Apathy towards farming in youth	✓	✓	✓	✓

15.5 PROPOSED INTERVENTIONS TO ENHANCE AGRICULTURAL PRODUCTION

In the light of inferences drawn out of SWOT Analysis and participatory evaluation of the agro-climatic and socioeconomic characteristics of the units, interventions under the following areas were recommended to be immediately implemented in order to enhance agricultural production in the units.

15.5.1 WATER CONSERVATION AND IRRIGATION

Community level organizational mechanism for better management of water involving farmers was recommended. Maintenance of canals and scheduling of irrigation water, through Water Users' Associations (WUAs) as envisaged in the "Kerala Irrigation and Water Conservation Act 2003" is suggested.

15.5.2 CROP PRODUCTION AND MANAGEMENT

All the four agro-ecological units selected, have favorable soil and climatic conditions to sustain diverse crop production systems. The lack luster performance of the crop sector could be attributed partly to the shortfall in achieving the basic infrastructure support like irrigation, seeds, manures, fertilizers, mechanization, storage and marketing. Production-system analysis showed that, rice continues as a less remunerative crop in the district compared to alternative crops like coconut, tapioca and banana (Rajalekshmi, 2006). Various schemes implemented so far in agriculture sector had only short-term objectives and therefore these could not address the sustainability of production systems. Policy interventions at microlevel planning that envisages horizontal and vertical integration among the panchayats and line department schemes for holistic development and infrastructure creation is the need of the hour. Being the rice granary of the state, special measures need to be taken for protecting and preserving the paddy fields. Considering the food security and ecological functions met by rice cultivation, a compensatory price mechanism has to be evolved to pay for the opportunity-cost of farmers in continuing rice cultivation and conserving the paddy fields.

A Food and nutrition security package for Palakkad district high-lighting the scope of bringing at least 50% of the summer rice fallows under arid and semi arid crops (pulses and grams) could be explored. Diversification and value-addition of coconut and its by- products on a commercial basis, organized efforts for tapping the potential of horticultural crops and their value addition and capacity building of the 'Kudumbasree' groups for undertaking enterprises by providing necessary forward and backward linkages including technology transfer are recommended.

15.5.3 LIVESTOCK PRODUCTION AND MANAGEMENT

Livestock has been an inevitable component in the integrated farming system followed in traditional farming in all the units. Even under hostile conditions, nearly 50% of the small/marginal farmers maintain at least one or two cattle. Suggestions for improvement include – specific interventions to evolve high yielding breeds of cattle suited to the climatic conditions, instituting easier and cheap loan facilities at nominal interest rate for longer period, establishing decentralized chilling facilities for storage of milk during rainy season and surplus production periods, establishing facilities for producing value added products of milk and capacity building and training to SHGs and providing the infra-structure facilities for processing and marketing and popularizing fodder cultivation and scientific fodder conservation like silaging and encouraging backyard poultry in homesteads.

15.5.4 RESEARCH AND EXTENSION STRATEGIES

The research strategy for the agro-ecological units under study should be the one addressing specific local problems. The units have reported that the yield of rice has been stagnant over the last few years under similar management level. The problems faced by agriculture in various units differed from each other. The research policy of the agricultural university needs to be restructured in such a way to facilitate demand-driven and location-specific research. These need to encompass climate change mitigation studies and strategies also.

The indices of technology adoption in major crops and dairying, fall in the range of 45- 60 in all the units. This implies that the recommended technologies and scientific management are not being fully adopted at the field level. SAARC Agricultural Vision 2020, has pinpointed the absence or weak Research-Extension-Farmer linkages resulting in large gaps in the farmers' practices and improved technologies. Effective revamping of the Research Extension interfaces between researchers and extension officials and the reorganization of field level agricultural offices may be fruitful. As the present organization of Krishibhavan, the field level implementing organ of the State Agricultural department, seldom give the Agricultural Officers enough time to involve in agricultural extension activities and the field problems, a change of the structural pattern of Krishibhavans creating the post of a clerical cadre to manage the office would be beneficial.

15.5.5 FORWARD AND BACK WARD LINKAGES

The constraints with regard to forward and backward linkages in farming include nonavailability of quality seeds and planting materials, nonavailability of fertilizers during peak cropping season, nonavailability of laborers and insufficiency of agro-machineries and lack of marketing facilities for meat and egg in the agro-ecological units. Establishment of seed villages and bio-control labs, decentralized storage and primary processing facilities at the padasekharam level, farmers' daily markets for vegetables and fruits, Labor Banks at panchayat levels, common facility centers may be done at appropriate locations, to provide adequate backward linkage and ensure hygienic production, standardization, packaging and labeling of agro-processing enterprises of Kudumbasree/SHGs in the agro-ecological units.

15.5.6 INSTITUTIONAL INTEGRATION AT GRASS ROOT LEVEL

Agricultural production system of any region comprises crop, live-stock and dairying, fisheries, irrigation, soil and water conservation. It is observed that, at the grass roots level, different agencies are functioning without any effective coordination and integration. Although irrigation is a crucial element in agricultural development, there is no mechanism at the grass-root level for scientific irrigation scheduling. In the case of finance and credit, the District Credit Plan is prepared without detailed consultation with stake-holders viz., farmers, elected representatives and extension officials of the concerned departments. Preparation of the Credit plans of banks at the grama panchayat level and its integration with Peoples' Plan Program would be more meaningful. It is evident that it is not the lack of resources and avenues, but the lack of comprehensive planning based on a perspective cutting across the boundaries of the departments, that is impeding sustainable growth and development. Integration of agricultural projects with major programs like MGNREGS for development of water sources, irrigation and drainage facilities, water harvesting structures, land development, horticulture development for SC/ST/BPL farmers and watershed development projects may be done. The unskilled labor force registered under MGNREGS should be organized to form labor banks to address labor scarcity during peak agricultural season. The labor budget and seasonality calendar under NREGs should be designed to suit the cropping season and peak agricultural operations of the panchayat, thus complementing agricultural production. The study strongly points out at the need for further efficient planning and allocation of resources viz., financial, physical and human resources at the grass root-level.

15.6 CONCLUSION

By virtue of its contribution to the district's domestic product and its dominant support for employment generation, farming and agri-related sectors need to be recognized as the key sector for providing livelihood security of Palakkad district of Kerala. Strategies devised on the basis of broad sectors or agro-ecological units or specific crops, should ultimately fall in line with the area development programs as supplementary and complementary programs, supported by local bodies. A spatial integration of crop, livestock and other agro-related enterprises under the aegis of Krishibhavans is suggested for the development of agriculture in any district. This requires integration of other line departments and major programs like MGNREGS in a participative manner with the support of local bodies. As a continuation of the present study, efforts to estimate the economics of the existing and the preferred systems and establishing value chains in agricultural commodities were suggested. Similar studies may have to be extended to all the districts of the state especially in a scenario of externalities like climate change and globalization.

KEYWORDS

- **Agro-Ecological Unit**
- **Fish Farms Development Agency**
- **Key informant farmers**
- **Kudumbasree Units**
- **National Rural Employment Guarantee Scheme**
- **Padasekara Samithis**
- **SWOT Analysis**
- **Water Users' Associations**

REFERENCES

Chambers, R. (1991). Shortcut and participatory methods for projects. Putting People First: Sociological variables in rural development (ed. Cernea, M.M), Oxford University Press, London, 31–76.

Economic Review (2010). State Planning Board, Government of Kerala.

Planning Commission, Government of India (2011). Report of the Working Group on Decentralized Planning in Agriculture for XII Plan.

Rajalekhsmi, P. (2006). Shifting of paddy cultivation in Palakkad district—An Economic Analysis. PhD thesis (unpublished), Dept. of Economics, University of Calicut.

www.agristat.com.

CHAPTER 16

CASHEW BASED CROPPING SYSTEMS: AN EFFECTIVE APPROACH FOR INTEGRATED LAND USE PLANNING IN HARD LATERITIC AREAS OF NORTHERN KERALA

A. V. MEERA MANJUSHA[1], AMBILI S. NAIR[2], and P. K. RETHEESH[3]

CONTENTS

[1]Assistant Professor (Horticulture), Regional Agricultural Research Station, Pilicode, Kerala Agricultural University, Pilicode (PO), Kasaragod (dist), Kerala

[2]Assistant Professor (Plant Breeding and Genetics), Regional Agricultural Research Station, Pattambi, Kerala Agricultural University, MelePattambi (PO), Palakkad (dist), Kerala

[3]Assistant Professor (Agronomy), Regional Agricultural Research Station, Pilicode, Kerala Agricultural University, Pilicode (PO), Kasaragod (dist), Kerala

ABSTRACT

The hard lateritic regimes of northern Kerala pose a serious threat to the agrarian economy of this landscape. In these areas no crop can be grown as the roots of most of the crops fail to break the hard laterite underneath, absorb water and survive. Hence these areas are often left barren. A survey in this area has revealed that with cashew as the main crop, such areas can be successfully brought into cultivation. Cashew thrives well in these hard laterites. Cashew (Anacardium occidentale L.) is a native of Latin American country, Brazil and is widely cultivated through out the tropics for its nuts. It was one of the few fruit trees from New World to be widely distributed through out the tropics by the early Portuguese and Spanish adventurers (Purseglove, 1988). Cashew has a long history as a useful plant but only in the present century it has become an important tropical tree crop. In Asia and Africa, small-scale local exploitation of cashew for its nuts and cashew apples started more than 300 years back. Cashew was introduced from Brazil into India by the Portuguese during the sixteenth century. Following its introduction into South Western India, the cashew probably diffused throughout the Indian subcontinent. Cochin served as a dispersal point for South East Asia as well (Johnson, 1973). It is presumed that the initial introductions in the Malabar Coast of Kerala were from only few trees and due to the hardy nature of the crop it has spread to all the coastal regions of India naturally. Initially it was introduced with the objective of preventing soil erosion and as a crop in eroded and marginal lands. It is this hardy nature of the plants that helps in its survival in the hard lateritic soils. In Northern Kerala while planting, farmers break open these laterites with mild explosives to make pits for planting cashew and fill the pit with topsoil and then planting is done. Alternatively soil is filled as a top layer in the entire plot up to certain height and cashew is planted. Once cashew is established it thrives well without any further soil amelioration requirements. Some farmers practice stone mulching with broken laterite pieces, which is found to effectively check the water loss during summer months. It is also assumed that the cashew roots aid in bringing amelioration of hardness in these laterites. A major threat faced by this landscape is the increasing laterization of the soil. If effective cropping systems with cashew as main crop could be developed more area could be made arable. With cashew, a crop that fetches high returns with minimal investment and little drudgery, the economically backward area of Northern Kerala could also be benefitted. Efficient technologies are also available for utilization cashew apple, which is otherwise treated as a waste product causing environmental problems. Further studies are warranted for refinement of cashew cultivation practices and the candidate crops in cropping system in these hard lateritic regimes as well as on effect of cashew root exudates on lateritic soils.

16.1 INTRODUCTION

The potential of trees to bring improvements in nutrition, income, housing, health, energy needs and environmental sustainability in the agricultural landscape has widely been acknowledged. Within the array of benefits brought by trees, an important element is the positive effect of trees on soil properties and consequently benefits for crops. This chapter explores current knowledge as to this relation between cashew and laterite soil, based on our experience and research. Laterite was first reported by Dr. Francis Hamilton Buchanan from Angadipuram in Kerala (India) on 20–21 December, 1800, while on a journey through the countries of Mysore, Canara and Malabar (Buchanan, 1807). Buchanan observed a type of weathered material used for building, which was an indurated clay with full of cavities and pores, containing a large quantity of iron in the form of red and yellow ochre. It was very soft when fresh and could be cut with any iron instrument. When exposed, it became hard and resisted air and water much better than any bricks. He coined the term – "Laterite" to designate this material. In Latin "laterite" means brickstone. However, there exists some dispute in regard to the authorship of the term "Laterite." Prescott is of the opinion that Babington (1821) was the first to use the term scientifically.

Laterite soils occupy an area of about 49,000 sq.miles in India. The laterite is specially well-developed on the summits of the Deccan Hills, Central India, Madhya Pradesh, the Rajmahal Hills, the Eastern Ghats, in certain plains of Orissa, Maharashtra, West Bengal, Kerala and Assam. These are found to develop under fair amount of rainfall and alternating wet and dry periods. The laterite and lateritic soils are characterized by a compact to vesicular mass in the subsoils horizons composed essentially of mixture of the hydrated oxides of aluminum and iron. These soils are deficient in phosphorus, potassium, calcium and magnesium. The pH is generally low. On higher levels these soils are exceedingly thin and gravelly, but on lower levels and in the valleys they range from heavy loam to clays and produce good crops, particularly rice. They are both in situ and sedimentary formations and are found all along the West Coast and also in some parts of the East Coast, where the rainfall is heavy and humid climate prevails. In the laterites on lower elevations paddy is grown, while tea, cinchona, rubber and coffee are grown on those situated on high elevations.

Lateritic soils are mostly climatogenic and vegetation and relief have played major roles in soil formation. In laterite areas, soil erosion is unabated during the monsoon aided by the landscape setting. Occurrences of plinthite at different depths, highly gravelly nature of subsoil, poor base status, low cation exchange capacity, moisture stress during summer coupled with low water table and presence of free oxides of Fe and Al are problems associated with their management. The laterite soils and some of the in situ developed red earths are highly exhausted soils and cannot sustain productivity on their own without supplemental fertilizers. In general, fertility status is medium to high in N, and low in available P and K. If proper fertilizer recommendations supplemented with organic manures are given, the production of coconut, cashew, mango, citrus, clove, nutmeg and cinnamon can be boosted in this soil. The most yield-limiting nutrient appears to be K followed by N (Chadha and Nair, 1998). The high-density planting, intercropping, checking erosion and biomass incorporation into the soil are highly necessary in this kind of soil.

In Kerala, between the western broad sea belt consisting of sandy and sandy loam soils and the eastern regions comprising the forest and plantation soils, the midland contains residual laterite. These are poor in total and available P_2O_5, K_2O and CaO. The nitrogen content varies from 0.03–0.33%; the lime is very poor and the magnesium is 0.11–0.45%. These soils are very low in bases, like calcium and magnesium, due to severe leaching and erosion. Due to low infiltration during rainy seasons, water stands above the hardpan, preventing growth of trees and other vegetation. During the dry season, the layer turns into a hard crust preventing root penetration and inhibiting plant growth.

16.2   METHODOLOGY

A survey was conducted in the lateritic areas of Northern Kerala to study the farmer's practices for standardizing ways to convert vast spread of hard lateritic regimes into cultivable lands. Generally these areas are left barren. People are averse to build dwelling units in these areas because of difficulties in breaking the hard pan. The places are often used for quarrying laterite blocks, which are used as a building material. After this the place is abandoned, so that water gets filled in these quarries and cause problems to inhabitants nearby. Year by year the quarried span is increasing. Environmental issues on account of quarrying and associated pollution problems are rising alarmingly.

16.3   RESULTS AND DISCUSSION

The survey revealed an interesting observation. Cashew formed the crux of any agricultural land use pattern. Cashew thrives well in these hard laterites. Cashew (Anacardium occidentale L.) a native of Brazil, is widely cultivated throughout the tropics for its nuts. It was one of the few fruit trees from New World to be widely distributed throughout the tropics by the early Portuguese and Spanish adventurers (Purseglove, 1988). Cashew has a long history as a useful plant but only in the present century it has become an important tropical tree crop. In Asia and Africa, small scale local exploitation of cashew for its nuts and cashew apples started more than 300 years back. Cashew was introduced from Brazil into India by the Portuguese during the sixteenth century. Following its introduction into South Western India, the cashew probably diffused throughout the Indian subcontinent. Cochin served as a dispersal point for South East Asia as well (Johnson, 1973). It is presumed that the initial introductions in the Malabar Coast of Kerala were from only few trees and due to the hardy nature of the crop it has spread to all the coastal regions of India naturally. Initially it was introduced with the objective of preventing soil erosion and as a crop in eroded and marginal lands. It is this hardy nature of the plants that helps in its survival in the hard lateritic soils. In Northern Kerala, while planting farmers break open these laterites with pick axes or mechanical drills or with mild explosives to make pits for planting cashew and fill the pit with top soil and then planting is done. Alternatively soil is filled as a top layer in the entire plot up to certain height and cashew is planted. Once cashew is established, it thrives well without any further soil amelioration. Some farmers practice stone mulching with broken laterite pieces, which is found to effectively check the water loss during summer months.

Cashew has deep and spreading root system. Root distribution pattern of cashew depends on factors such as age of the tree, type of planting material, soil environment in which it is grown, level of nutrition and irrigation. Majority of the feeding roots are present in the surface layer of the soil. The root penetration studies in cashew indicate that it is a surface feeder. Nutrient absorption was mostly from the 0 to 15 cm soil layer than from the deeper zone. The lateral as well as vertical spread of the roots was determined on dry weight basis by Khader (1986). The results revealed that 97.87% of thicker roots and 81.27% of fine roots were spread over within a radius of 2 m of the tree and 90.84% of thicker and 53.74% of fine roots were observed from 0–1 m depth. The maximum depth up to which the cashew roots penetrated was 9.5 m. Studies using radio isotope ^{32}P soil-injection technique, indicate distribution pattern of active roots of cashew up to a distance of 4 m from the tree and to a soil depth of 60 cm. The results designate that active roots are confining to the top 15 cm of soil layer. About 72% of the root activity was found within a radial distance of 2 m from the tree (Wahid et al., 1989). Hence, in lateritic soils with a hard pan beneath, breaking open the hardpan with pick axes or mechanical drills or with mild explosives is successful on account of the fact that in the active feeding layer, the soil is loose, penetrable and acts as the source of nutrients for growth. Once the crop gets established, the roots penetrate to greater depths breaking the hard pan and provide anchorage to the tree. It is this peculiarity of the crop that makes it ideal for eroded and marginal lands.

Jobbagy and Jackson (2001) found that cycling mediated by plants exerts a marked influence on the vertical distribution of nutrients in the soil, especially in the case of more limiting nutrients such as P and K which is specially the case with laterite soils. Patterns of greater concentration of these nutrients in surface layers (0–20 cm) were attributed to the fact that since these are more important to plants, they are subject to greater uptake and cycling, being absorbed from deeper layers and returned to the soil surface through litterfall and rain water throughfall. This process of uptake functions in opposition to leaching, which moves nutrients downward and acts more strongly on those nutrients that are in less demand by plants. If a nutrient is not limiting, its movement in the soil profile will be more influenced by leaching than by cycling and it will be present in higher concentrations at greater depth, as occurs with Na, Cl and Mg (Jobbagy and Jackson, 2001). Ulery et al.

(1995) found this sort of pattern in soils influenced by the presence of four planted tree species with increments of almost three times as much K in the surface layer in relation to the original soil before planting, while below 20 cm this increment was absent or negative. These studies suggest that the limiting nutrients P and K of laterite soil will be more concentrated in surface layer, which is advantageous for the surface feeders like cashew. Their study also showed a high degree of leaching of Na, which is less in demand by plants. The presence of higher Na content in deeper layer helps in deflocculation that will aid in breakage of hard lateritic pan present beneath the soil.

Canopy biomass fallout of leaves, cashew apples and flowers and the subsequent nutrient release was calculated to supply 15.5–37.7% of tree total requirements of macronutrients at sixth year for cashew (Richards, 1992). This may also be one of reasons for successful establishment of cashew in laterites though it is marginal in nutrient content.

Rao (1987) estimated the interception losses of precipitation from cashew in humid tropical region of Kottamparamba, Kerala, India. The storage capacity of the Cashew trees was worked out as 0.8 mm and the throughfall coefficient as 0.391. The trees under observation were 15–20 years of age with a leaf area index of 1.0–1.25. About 31% of the storm rainfall for storms of 25.0 mm was intercepted by the Cashew trees and lost to the atmosphere. The interception values for cashew is fairly high indicating that cashew is able to reduce the precipitation intensity and this interception washes solid particles and dissolved carbon from leaves affecting soil and water chemistry and weathering processes, which are of great consequence in the high rainfall area affecting the formation of hard pan in the soil.

Nair et al. (2009) found values for soil carbon stock in various ecosystems, which revealed a general trend of increasing soil carbon sequestration in agroforestry systems when compared with other land use practices other than forestry. A study by Aweto and Ishola (1994) found that the levels of organic carbon, nitrogen, exchangeable calcium and magnesium, and available phosphorus were similar under logged forest and cashew, suggesting that organic matter and nutrient cycles in a cashew plantation are similar to those in a logged rain forest. This merits cashew based cropping systems for consideration as tool for mitigation of climate change brought about by Greenhouse gas emissions.

The role of rhizosphere-associated microorganisms is crucial in changing the physical and chemical properties of soil under cultivation. Little has been studied about the rhizosphere microbes of cashew and their role in ameliorating the soil properties in lateritic areas. Krishnaraj and Gowda (1990) have reported the presence of phosphate solubilizing bacteria in the endorhizosphere to an extent of $0.21 \pm 0.4 \times 10^6$/g root in cashew. This assumes greater significance that unlike rhizosphere and soil microorganisms, endorhizosphere microorganisms are closely associated within the plants with greater degree of specificity.

Due to unimodal pattern of rainfall resulting in long dry spell, especially during the critical period of flowering and fruit set, cashew suffers from severe physiological stress in the lateritic soils of Kerala. The hardiness and adaptability to water stress conditions of cashew were attributed mainly to the higher photosynthetic rates even at full irradiance and high vapor pressure deficit (>2.5 kPa). The CO_2 compensation concentration was observed to be at about 80–100 cm^3/m^3. The intercellular ambient ratio of CO_2 was 0.8–0.86 based on which cashew was classified under C3 species. These results indicated the preference of cashew trees for the place of abundant irradiance (Balasimha, 1991). Latha and Abdul Salam (2003) reported that root:shoot ratio remained unaffected up to 90 days with depletion of available water, indicating the capacity to withstand varied degree of soil moisture stress. These physiological features might have helped for better survival of cashew in the adverse conditions in lateritic soils. More studies in this area are warranted as understanding the physiological responses in this condition will supplement the high yielding varieties thereby increasing production and productivity.

Kannur and Kasaragod districts of Kerala produce the premium quality cashew nuts in the world itself. Cashew is crop, which gives fairly high returns with minimal investments. The labor requirement for the crop is very low except for harvesting. If effective cropping systems with cashew as main crop could be developed more area could be made arable. With cashew, a crop that fetches high returns with minimal investment and little drudgery, the economically backward area of Northern Kerala could also be benefitted. Efficient technologies are also available for utilization of cashew apple, which is otherwise treated as a waste product causing environmental problems. Further studies are warranted for refinement of cashew cultivation practices and the candidate crops in cropping system in these hard lateritic regimes, changes in soil physical, chemical and microbiological properties brought about by cashew based cropping systems as well as on effect of cashew root exudates on weathering and amelioration of lateritic soils.

KEYWORDS

- **Amelioration**
- **Cashew**
- **Endorhizosphere Microorganisms**
- **Harvesting**
- **Lateritic Pan**

REFERENCES

Aweto, A. O., Ishola M. A., (1994). The impact of Cashew (**Anacardium occidentale**) on forest soil. Exp. Agric., 30, 337–341.

Babington, B. (1821). Remarks on the geology of the country between Tellicherry and Madras. Trans.Geol.Soc. Lond., 5, 328–339.

Balasimha, D. (1991). Photosynthetic characteristics of cashew trees. Photosynthetica. 25, 419 – 423.

Buchanan, F. (1807). A journey from Madras through the countries of Mysore, Kanara and Malabar, East India Co., London, 2, 436–460.

Chadha, K. L., Nair, M. K. (1998). Red and lateritic soils. Volume 1: Managing red and lateritic soils for sustainable agriculture. Sehgal, J., Blum, W. E., Gajbhiye, K. S. (Eds), 179–189.

Jobbagy E. G., Jackson, R. B. (2001). The distribution of soil nutrients with depth: global patterns and the imprint of plants, Biogeochemistry, 53:51–77.

Johnson, D. (1973). The botany, origin and spread of cashew (Anacardium occidentale L.). J. Plant. Crops, 1, 1–7.

Khader, K. B. A. (1986). Distribution of cashew roots in the laterite soils of West coast of India. Ind. Cashew J., 17(2), 15–17.

Krishnaraj P. U., Gowda, T. K. S. (1990). Occurrence of phosphate solubilizing bacteria in the endorhizosphere of crop plants. Curr Sci., 59: 933–934.

Latha, A., Abdul Salam, S. (2003). Response of cashew seedling to soil moisture stress. The Cashew, 17: 17–21.

Nair, P. K. R., Kumar, B. M., Nair, V. D. (2009). Agroforestry as a strategy for carbon sequestration, J. Plant Nutr. Soil Sci., 172:10–23.

Purseglove, J. W. (1988). Tropical Crops- Dicotyledons. Longman, UK.

Rao, A. S. (1987). Interception losses of rainfall from cashew trees. J. Hydrol., 90: 293–301.

Richards, N. K. (1992). Composition and Nutrient Cycling Sandy Red Earths of Northern Territory Australia. Sci. Hort., 52, 124–142.

Ulery, A. L., Graham, R. C., Chadwick, O. A. Wood, H. B., (1995). Decade-scale changes of soil carbon, nitrogen and exchangeable cations under chaparral and pine, Geoderma, 65:121–134.

CHAPTER 17

RURAL AND AGRICULTURAL DEVELOPMENT THROUGH SOIL AND WATER CONSERVATION: HARIYALI PROJECT OF VARAKKAD WATERSHED OF NILESHWAR BLOCK, KASARAGOD DISTRICT, KERALA STATE

M. J. MERCYKUTTY[1]

CONTENTS

[1]Associate Professor, College of Agriculture, Padannakkad, Kasaragod (Dist.), Kerala.

17.1   INTRODUCTION

In India in the last three decades, watersheds have become the pivotal unit for rural development programs. Growing concern of poverty, population growth and environmental degradation have led to increasing public investment in India towards integrating resources through watershed management. The term 'watershed development' has ideally been accepted "as a geophysical unit for planning and executing development program for rational utilization of all natural resources for sustained optimum production of biomass with the least damage to the environment" (GoI, 1999:ix). However, due to high importance placed to about 53% of the total geographical area subjected to degradation, soil and water conservation measures have been adopted towards improving agriculture production and productivity to cater to large percentage (70%) of rural population. Since 1994, the country has wide variety of experiences in facilitating the watershed program. Eventually, the Guidelines of watershed development program are revised in 2001 (Watershed Guidelines – Revised) and 2003 (Hariyali). Apart from these guidelines, Ministry of Agriculture also issued guidelines for National Watershed Development Project for Rainfed Areas (2000). The potential of Watershed becoming a functional unit remains in understanding the linkages and promoting economic activities that sustain watershed management.

17.2   PARTICIPATORY WATERSHED MANAGEMENT

The key to the success of any watershed project and its sustainability depends on people's participation. For achieving the desired participation of people, the roles of community organizations, groups and other stakeholders are crucial. Local people must play an active role starting from project design, moving to implementation and the project maintenance. In this context, a participatory watershed management approach is considered as the ideal for achieving food security and sustainability (Budumuru Yoganand et al., 2006). People's KNOWLEDGE and SKILLS must be seen as a potentially positive contribution to the project. A participatory project should seek every possibility to base its activities upon local resources, both to avoid situations of dependence on external ones and also to help develop local capabilities, which will be important if the development is to be sustained.

The results of the participatory research carried out in the Amachal watershed show that incorporating farmers in an innovation process helps them to address their own problems as well as seek appropriate information when necessary. Also the participatory approach enables the community to visualize and evaluate the impact of innovative technologies (Shubha Vishnudas, 2006). It has been noted that participatory watershed management projects have been raising income, agricultural productivity, generating employment and conserving soil and water resources (Vidula et al., 2012).

17.3   HARIYALI PROJECT OF VARAKKAD WATERSHED

In Nileshwar block of Kasaragod Dt, Kerala State, Hariyali project is being implemented in 11 water sheds at a total area of 5900 Hectors distributed in eight panchayats from 2007 onwards. The total outlay of the project for Nileshwar block is Rs.3.54 Crores. The technical support of the Hariyali project was undertaken by the College of Agriculture, Padannakkad, Kasaragod Dt of Kerala Agricultural University

The primary objective of the Hariyali project is harvesting every drop of water and improving the status of the rural farmers. Activities for harvesting every drop of rainwater for the purpose of irrigation including horticulture and floriculture, pasture development, fisheries, etc., as well as for drinking water supply are envisaged. Emphasis is also given on employment generation, poverty alleviation, community empowerment and development of human and other economic resources of the rural areas. It also aims for mitigating drought and desertification of crops, human and livestock population for over all improvement of rural areas. Steps are also ensured for conservation and development of natural resources and improving vegetative covers, promoting use of simple easy and affordable technological solutions and institutional arrangements that of build up on local technical knowledge and available materials.

This watershed management approach is participatory in nature; people friendly, location specific, process based and geared to cater to the problems and needs of the rural communities. This study was conducted in Varakkad watershed of West Eleri panchayath of Nileshwar block, which comprised of ward II, III, VIII, IX, X, XI, and XII. The watershed has a total area of 1085. 54 Hectare. Eventhough the treatable area of the watershed is 900 Hectare, there is a watershed project already executed under RIDF of NABARD with an area of 275 Ha. Two hundred respondents were covered under the study using proportionate stratified random sampling, with the wards as a basic unit of stratum. The study envisaged to analyze the personal, sociocultural and techno-economic factors of the respondents in the area of the study. Apart from Agriculture and resource survey different soil conservation measures were also analyzed. Various tools viz; Participatory Rural Appraisal Techniques, focused group interviews, surveys, field observations, etc., were employed for the study. To ensure optimum and sustained productivity through scientific planning, the watershed needs a decision making information system that involves an appraisal of agro-ecological characteristics, resource limitations and potential of the watershed for resource development. This complete information helps in generating an information system for watershed management.

17.4   GENERAL DESCRIPTION OF THE WATERSHED

The watershed area is mainly having the crops such as rubber, coconut, areca nut, and banana. The main crop cultivated is rubber and there is also mixed cropping of coconut and banana, coconut-areca nut and banana is also seen.

Regarding general socioeconomic situation, majority of families come under BPL, most of the people are farmers. Twenty percent are doing nonagricultural activities.

Geographical description of the watershed is slope to plain with laterite type of soil. The area was highly prone to soil erosion. Due to the continuous rains soil was eroded yearly and hence protection of soil from erosion was inevitable. The streams had to be protected by constructing sidewalls. Contour bunds should be constructed across the sloppy area to prevent soil erosion. Low productivity of the crops was the major factor. The area was highly prone to soil erosion during rainy days and nonavailability of water during summer months. On closer observation it was found that soil and water conservation measures were not adopted properly and streams were not well protected. All these factors resulted in leaching of nutrients along with soil, which in turn reduces productivity of the crops. Moreover the streams of the area are having a high velocity of flow of water during rainy season. During summer months water is not flowing through the streams, which results in scarcity of water for irrigation.

The findings of the sociocultural and techno-economic factors of the respondents are discussed below.

TABLE 17.1   Distribution of Respondents Based on Gender

Category	Ward II		Ward III		Ward VIII		Ward IX		Ward X		Ward XI		Ward XII		Total	
	N	%	N	%	N	%	N	%	N	%	N	%	N	%	N	%
Male	31	47	159	56	19	51	66	43	21	55	33	59	54	52	383	52
Female	35	53	127	44	18	49	86	57	17	45	23	41	50	48	356	48
TOTAL	66	100	286	100	37	100	152	100	38	100	56	100	104	100	739	100

A perusal of Table 17.1 reveals that, 52% of the total respondents were male and 48% were female. Ward wise analysis showed that ward II and IX have slightly more female population while other wards have slightly more male population.

TABLE 17.2   Distribution of Respondents Based on APL/BPL Ratio

Category	Ward II		Ward III		Ward VIII		Ward IX		Ward X		Ward XI		Ward XII		Total	
	N	%	N	%	N	%	N	%	N	%	N	%	N	%	N	%
BPL	9	37.5	14	17	1	10	10	25	1	14	6	46	8	33	49	24.5
APL	9	37.5	57	70	8	80	23	57.5	4	57	6	46	15	63	122	61
NR	6	25	11	13	1	10	7	17.5	2	29	1	8	1	4	29	14.5
TOTAL	24	100	82	100	10	100	40	100	7	100	13	100	24	100	200	100

It is evident from Table 17.2 that 61% of the total population in the area belonged to APL category and (24.5%) belonged to BPL category. The remaining respondents did not respond. Viewing the ward wise distribution, it is interesting to note that the BPL ratio and APL almost equal in ward II and XI (37.5 and 46%, respectively) and highest percentage of BPL was in ward XI (46%).

TABLE 17.3 Distribution of Respondents Based on Proper Awareness on Soil Conservation

Category	Ward II		Ward III		Ward VIII		Ward IX		Ward X		Ward XI		Ward XII		Total	
	N	%	N	%	N	%	N	%	N	%	N	%	N	%	N	%
Yes	9	37.5	2	2	0	0	1	2.5	0	0	0	0	15	63	27	13.5
No	11	45.8	62	76	9	90	14	35.0	0	0	0	0	0	0	96	48
NR	4	17	18	22	1	10	25	62.5	7	100	13	100	9	38	77	38.5
TOTAL	24	100	82	100	10	100	40	100	7	100	13	100	24	100	200	100

It is observed from Table 17.3 that only 13.5% of respondents are aware of soil conservation measures. However, respondents of ward VIII, X, and XI perceived that they did not need soil conservation measures. Forty eight percent of total respondents were not aware and do not need soil conservation method and 38.5% of the total respondents did not express any opinion out of which 100% of responders were from ward X and XI.

TABLE 17.4 Distribution of Respondents Based on Market utilization

Category	Ward II		Ward III		Ward VIII		Ward IX		Ward X		Ward XI		Ward XII		Total	
	N	%	N	%	N	%	N	%	N	%	N	%	N	%	N	%
Yes	16	66.7	73	89	10	100	28	70	4	57	11	85	20	83	162	81
No	1	4.2	5	6	0	0	7	17.5	0	0	1	8	0	0	14	7
NR	6	25	4	5	0	0	5	12.5	3	43	1	8	4	17	23	11.5
TOTAL	24	100	82	100	10	100	40	100	7	100	13	100	24	100	200	100

Table 17.4 indicates that 81% of the respondents were utilizing market facilities. The result of ward wise analysis was not on par with the total. Ward VIII showed distinctly higher percentage (100%) of respondents utilizing market facilities, where as, in case of wards X, it was only 57%. Seven percent of the respondents were not utilizing market facilities and 11.5% of total respondents did not express any opinion.

TABLE 17.5   Distribution of Respondents Based on Credit

Category	Ward II		Ward III		Ward VIII		Ward IX		Ward X		Ward XI		Ward XII		Total	
	N	%	N	%	N	%	N	%	N	%	N	%	N	%	N	%
Yes	12	50.0	35	43	5	50	20	50	7	100	5	38	15	63	99	49.5
No	0	0.0	0	0	0	0	7	17.5	0	0	0	0	0	0	7	3.5
NR	12	50	47	57	5	50	13	32.5	0	0	8	62	9	38	94	47
TOTAL	24	100	82	100	10	100	40	100	7	100	13	100	24	100	200	100

From Table 17.5, it could be observed that only 49.5% of total respondents were using credit while ward wise distribution showed that the highest percentage was for ward X and XII (100% and 63%, respectively) and 47% people did not express any opinion.

TABLE 17.6   Distribution of Respondents Based on Labor Utilization

Category	Ward II		Ward III		Ward VIII		Ward IX		Ward X		Ward XI		Ward XII		Total	
	N	%	N	%	N	%	N	%	N	%	N	%	N	%	N	%
Yes	3	12.5	5	6	0	0	2	5	2	29	1	8	17	71	30	15
No	20	83.3	73	89	0	0	14	35.0	0	0	0	0	0	0	107	53.5
NR	1	4	4	5	10	100	24	60.0	5	71	12	92	7	29	63	31.5
TOTAL	24	100	82	100	10	100	40	100	7	100	13	100	24	100	200	100

A perusal of Table 17.6 revealed that on an average only 15% of total respondents were utilizing laborers. Among the respondents, 53.5% were not utilizing the labor and 31.5% of the total respondents did not express any opinion in the above parameters.

TABLE 17.7   Distribution of Respondents Based on Skilled Labor Availability

Category	Ward II		Ward III		Ward VIII		Ward IX		Ward X		Ward XI		Ward XII		Total	
	N	%	N	%	N	%	N	%	N	%	N	%	N	%	N	%
Yes	3	12.5	5	6	0	0	2	5	2	29	0	0	17	71	29	14.5
No	20	83.3	73	89	0	0	14	35.0	0	0	0	0	0	0	107	53.5
NR	1	4	4	5	10	100	24	60.0	5	71	13	100	7	29	64	32
TOTAL	24	100	82	100	10	100	40	100	7	100	13	100	24	100	200	100

Analysis of Table 17.7 indicates that, only 14.5% of the total respondents have the availability of skilled labors. Though the results of wards XII was found to be on par with the total result 71% and ward VIII and XI showed that skilled laborers are not available around 53.5% also do not have readily available skilled labor and 32% of the total respondents did not express any opinion on this.

TABLE 17.8 Distribution of Respondents Based on Drainage

Category	Ward II		Ward III		Ward VIII		Ward IX		Ward X		Ward XI		Ward XII		Total	
	N	%	N	%	N	%	N	%	N	%	N	%	N	%	N	%
Yes	0	0.0	1	1	0	0	2	5	0	0	0	0	0	0	3	1.5
No	20	83.3	65	79	9	90	13	32.5	0	0	0	0	0	0	107	53.5
NR	4	17	16	20	1	10	25	62.5	7	100	13	100	24	100	90	45
TOTAL	24	100	82	100	10	100	40	100	7	100	13	100	24	100	200	100

Table 17.8 shows that the respondents utilizing drainage measures were only 1.5%. Ward wise analysis result was found to be on par with the total. Ward VIII, X, XI and XII showed that respondents did not use drainage measures. Among the respondents, 45% did not respond to the above said parameters.

17.5 CONSERVATION BASED INTERVENTIONS

Considering the gravity of soil erosion and other prevailing features of the area following works were suggested and carried out. In case of Arable land conservation the measures are Stone pitched contour bunding, Agrostological measures, Bit Trenches/Moisture conservation pits, Centripetal terraces and Agroforestry planting. Non-Arable land conservation measures are Retaining wall, Loose boulder check dams, Water Harvesting Structures, Silt Collection Tanks and Shuttering for existing V.C.Bs. A brief description of these measures are followed.

17.6 STONE PITCHED CONTOUR BUNDS

They consist of building earthen enhancement across the slope of the land, following the contour as closely as possible. A series of contour bunds divide the area into strips and act as barriers to the flow of water, thus reducing the amount and velocity of runoff. When bunds are constructed strictly on contour lines, it will hold the entire water coming from the interspaces between two successive bunds and the water thus stored will gradually gets infiltrated into the soil, which ultimately help in replenishing ground water storage. In addition, the eroded soil will be deposited behind the bund. Bunds with granite pitching are suitable for more-sloppy areas with heavy runoff.

17.7   AGROSTOLOGICAL MEASURES

It was proposed to strengthen the constructed stone pitched contour bund by growing fodder grass over it. This will not only strengthen the bunds by its fibrous root system but also give food for the livestock. Hence the animal husbandry of the watershed area can be increased.

17.7.1   BIT TRENCHES/MOISTURE CONSERVATION PITS

Moisture conservation pits play a major role in soil and water conservation. These pits have to be dug out in the middle and valley portions so as to collect the down flowing water and silt. The water thus collected will percolate in the soil and increases water availability in the down stream. The silt thus collected, which is fertile topsoils eroded from upper streams, can be used to the neighboring crops (Table 17.9).

TABLE 17.9   An Over View of Major Works Undertaken Arable Land Treatments

Works	Quantity		
	Phase I	Phase II	Phase III
Vegetative barriers	1900 Rm	10,000 Rm	5100Rm
Stone Pitched Contour bunds	1900 m2	10,000 m2	5100m2
Centripetal terraces/circular trenches	5000 Nos	5000 Nos	4000 Nos
Bit trenches/moisture conservation		6000 Nos	6000 Nos
Afforestation works (Tree planting)		5500 Nos	7000 Nos

17.7.2   CENTRIPETAL TERRACES/CIRCULAR TRENCHES

Circular trenches which are dug out around perennial plants conserve moisture in-situ.

17.7.3   AGROFORESTRY PLANTING

Agroforestry saplings and forestry saplings, which have timber value, can be included in this, which help in minimizing the soil erosion and also conserve moisture. Besides, microclimate of the area will be changed by the canopy of the forest saplings. Plants such as mango, jack, neem, subabul, mahagony, teak, etc., can be promoted.

17.7.4   RETAINING WALLS

The banks of streams and gullies within the watershed can be protected by constructing retaining walls. Establishing vegetation such as pandanus, vetiver and glyricidia along stream banks also proved to be effective for protecting the banks from the erosive action of water flowing through the stream.

17.7.5   LOOSE BOULDER CHECK DAMS

They reduce the gully bed slope thereby reducing the velocity of runoff water, preventing the erosion and down cutting of gully beds. This type of check dam is fit for comparatively deeper streams. It can be constructed with locally available boulders and stones with an average height of 60–90 cm (Table 17.10). The center of the dam is kept lower than the sides to form the spillway. A chain of loose boulder check dams can reduce the runoff velocity and aid in the recharging of water.

TABLE 17.10   Non Arable Land Conservation Measures

S. No.	Details of Work	Quantity
1	Retaining wall	700 Rm
	A type (1 m height)	
	B type (1.5 m height)	400 Rm
2	Loose boulder check dams	100 Nos
	Type A (1 m width)	
	Type B (1.5 m width)	80 Nos
	Type C (2 m width)	50 Nos
3	Water Harvesting Structure	2 Nos
	$5 \times 5 \times 3$ m3	
4	Silt Collection Tank	1 No
5	Shuttering existing V.C. Bs	3 Nos
6	Self Help Groups	10 Nos

17.7.6   SILT COLLECTION DEVICES/PONDS

Ponds are the common structures used for rainwater harvesting. They are very important in using rainwater for drinking, irrigation purposes and for recharging ground water storage. New ponds can be constructed on sides of the drainage line. Side protection works such as rubble walls/stone pitching, etc., can also be done.

The existing ponds can be developed by desilting or deepening and strengthening side with vegetative or structural measures.

17.7.7 VENTED CROSS BAR (VCB)

The most common type of check dam constructed by the governmental agencies is VCB. The construction cost of VCB is comparatively lesser than that of concrete dams. The vents are closed by putting wooden planks in two layers and earth is filled in between them. A well tamped earth fill is made against the upstream face of the head wall and brought to the height of the spillway crest. Generally VCB is plugged after the south-west monsoon and opened again before the onset of monsoon. It is an effective structure to control runoff and collect water

In brief, these different interventions will be very effective to control soil erosion and to conserve water, which will be used by many families for drinking and for irrigation purpose. Through Shutter provision to VCB water can be blocked during summer month, which will facilitate the infiltration, and it will also help the people to cultivate more crops with proper irrigation. These Watersheds development programs will also result in bringing several positive trends including diversification of the rural economy, development of new institutions, increasing cropping intensity, improved fodder production, capacity development of the community, etc.

17.8 CONCLUSION

The impact of watershed is multifaceted. In the case of agricultural productivity we find that first the changes take place in the land use pattern, which is visible by an increase in area brought under cultivation and also by bringing the marginal lands under plow. Second, by changing the cropping pattern on the marginal lands from low density-low value crops to horti-silvi pastoral or regular crop system. Third, direct impact on the production and productivity of the crops are the significant changes marking a step towards a sustainability of the technology used in cultivation of the crops. The productivity increases for traditional as well as commercial crops. In the socioeconomic sphere, watershed enhances the incomes of the people through rise in productivity, employment opportunities and rise in wages. The income rise influences the asset formation, rise in expenditure on education, health and growth of horticulture and nonfarm sector. These in turn influences the socioeconomic well being of the people in terms of reduction of poverty, rise in standards of education and access to health facilities, etc. Further, the availability of the infrastructure like transport, health and communications and other agricultural extension services also accelerate the performance of Agricultural sector (Deshpande and Narayanamoorthy, 1999).

Since the physical benefits of watershed program are quite vividly apparent maintenance of watershed structures after the withdrawal of government support is quite crucial for the sustainability and success. One of the major impacts of the watershed program is that it enhanced a general awareness about the various policies and programs, which are implemented at Panchayat level.

Though some amount of physical benefits could be accrued through watershed implementation quite easily; however, it would take a much longer time and sustained efforts to realize the participatory and other holistic benefits of watershed program.

In nutshell, the impact of watershed management depends on effectiveness of the technology in the background of needs, priorities, cultural practices and community participation.

KEYWORDS

- **Amachal Watershed**
- **Horti-Silvi Pastoral**
- **Varakkad Watershed**
- **Watershed Development**

REFERENCES

Budumuru Yoganand, Tesfa G., Gebremedhin, (2006). Participatory Watershed Management for sustainable rural livelihoods in India||, Research paper 2006.

Deshpandy, R. S., Narayanamoorthy, A. (1999). 'An Appraisal of watershed development program across regions of India,' Artha vijnana VolXLI, No.4, December.

Government of India (2003). Guidelines for Hariyali, Ministry of Rural Development, Government of India.

Government of India (GoI). (1999). National Commission for Integrated Water Resources Development Plan.Ministry of Water Resources, New Delhi. September.

Government of India (2000). Common Approach for Watershed Development, New Delhi, Ministry of Agriculture and Cooperation.

Ministry of Rural Development, Government of India (1994). Guidelines for Watershed Development.

Ministry of Rural Development, Government of India (2001). Guidelines for Watershed Development – (Revised 2001).

Shubha Vishnudas, (2006). Sustainable watershed management Illusion or reality? A case of Kerala state of India, 2006.

Vidula A. S., Kulkarni, S. S., Santosh Kumbhar, Vishal Kumbhar (2012). Participatory Watershed Management in South Asia: A Comparative Evaluation with Special References to India, International Journal of Scientific & Engineering Research, Vol 3(3).

WASSAN (2003). Hariyali – Issues and Concerns.

PART IV

INTEGRATED LAND USE PLANNING AND INSTITUTIONAL
ARRANGEMENTS

CHAPTER 18

INTEGRATED LAND USE PLANNING—A KNOWLEDGE AND DECISION SUPPORT SYSTEM FOR SUSTAINABLE AGRICULTURAL AND RURAL DEVELOPMENT AT BLOCK LEVEL

M. MONI[1]

CONTENTS

[1]Deputy Director General, (Agricultural Informatics), National Informatics Centre, Government of India, New Delhi

ABSTRACT

Agricultural Development and Rural Development are interdependent at grassroots level (i.e., block level), for agricultural productions (input, cropping, output and postharvest system). However, during the last three decades, gap has been widening. There are Government Institutional delivery system is operational, at block (i.e., numbering about 6500 in India), for agricultural services and rural development services and these are mostly related to developmental schemes. The officers associated with these two vital rural economy sectors, about 6500 agricultural development officers and about 6500 rural development officers, are not equipped with land resources management system capabilities, in general and agricultural resources management system capabilities in particular. Land Use Planning, as a Knowledge and Decision Support System is not existent as a matter of essential requirement, for sustainable development. Block level Administration has not been equipped with Agricultural Resources Information System (AgRIS) in digital mode.

This chapter deals with the need for undertaking capacity building for about 13,000 agricultural and rural development officers on Land Use Planning, and also involvement of Geography discipline from Schools and Colleges of the Block, for Natural Resource Management (NRM), and facilitate advocacy through Information service of Agriculture Mission Mode e-Governance Program. The District Annual Plan outlay, which amounts to an average of Rs. 1200 Crores, requires UTILISATION based database based micro level planning. Information Service on "Integrated Land Use Planning for Sustainable Agricultural and Rural Development" is to be introduced as a SERVICE under the Agriculture Mission Mode Project (AMMP), in collaboration with NBSS&LUP and State Land Use Boards.

18.1   PROBLEM BASE-ANALYSIS—AN INTRODUCTION

The declining importance of agriculture in India, about 15% of the GDP, is a normal transformation accompanying economic development. Still, crowding of workers takes place in the Agricultural sector. The National Food Security Bill 2009, which envisages to cover about 65% of the Indian population, forcibly require agricultural development and rural development in a sustainable manner, when the agricultural land have been getting converted into nonagricultural purposes. The Small and Marginal Farmers, who constitute about 82% of the Indian farming community, are the backbone of the food security net, and facing problems associated with production, productivity, inputs, storage, processing, marketing, extension services, etc. Ecodegradation, environmental hazards and limited natural resource base are the challenges faced by small and marginal farmers. On-Farm S&T Practices, Off-Farm S&T practices and Non-Farm S&T practices are not being provided with required thrust and support. Farm based databases and Farmer based databases (agricultural statistics) are nonexistent at grassroots level. With natural resources becoming scarce, application of integrated Land Use Planning for sustainable agricultural and rural development is called for, as a KDSS at block level.

Agricultural sector is fully endowed with, only ill-structured and semistructured problems to be solved and hence adoption of decision support systems (DSSs) is required, based on database technology, statistical techniques, management sciences, knowledge bases, expert systems, AI techniques, web technology, internet technology, multimedia technology, software technology, mobile technology (3G, 4G), GIS technology and remote sensing technology, content management, modeling and simulation, forecasting, precision engineering (wireless and sensors), data/knowledge mining, knowledge discovery, etc. It is essential to leverage on a mix of emerging and existing technologies for effective and inexpensive ICT penetration for agricultural development of the Country. Geography/Biology Discipline at School levels and College levels, and Computer Sciences Discipline at College levels in blocks are needed to be associated on a continuous basis.

Agricultural Research Systems, Input Systems, Production Systems and Output System require to be built-in with effective ICT enabled "Information Systems," capable of delivering services, in Indian local languages, for enhancing agricultural production, productivity and income rise, decision making, planning, monitoring and intervention in the interest of all stakeholders. Agricultural Informatics has emerged as a discipline, out of Synergization between Computer Science and Technology and Agricultural Science and Technology.

Indian Agricultural sector has begun its digital journey since 1987 with the launching of India's ambitious District Information System (DISNIC) by National Informatics Centre (NIC), in 520+ districts of India, wherein the DISNIC-AGRIS (Agricultural information System) project was one among the 27 sectoral database projects with "Village as its basic Unit." Major Blueprint on "agricultural informatics" viz., AGMARKNET, AGRISNET, SeedNet, PPIN, FISHNET, APHNET, etc., was made available to the Country through a National Conference on "Informatics for Sustainable Agricultural Development (ISDA-95) held in May 1995.

National IT task Force (1999) also recommended, vide its recommendation no. 79, that "the Government shall take necessary steps to boost IT for Agriculture and Integrated Rural development." In 2006, the Government of India announced its National e-Governance Program (NeGP) which included "Agricultural Mission Mode Project" for providing information services to farming community through mobile phones, IVRS, email, web-enabled access in Indian languages. Agricultural Informatics, as a discipline, has emerged out, to usher in agricultural dynamism in India.

The Comprehensive Agricultural Development Plan (CDAP) has not visualized Natural Resources Management (NRM) and Land Use Planning (LUP) as its main ingredients. Many CDAPs are yet to be operationalized.

18.2   CHALLENGES AND OPPORTUNITIES IN NATURAL RESOURCES MANAGEMENT

NRM refers to the management of natural resources such as land, water, soil, plants and animals, with a particular focus on how management affects the quality of life for both present and future generations. NRM is congruent with the concept of sustainable development, a scientific principle that forms a basis for sustainable global land management and environmental governance to conserve and preserve natural resources.

Rainfed agriculture is crucial to our country's economy and food security because 60% of the net sown area is Rainfed and about 44% of the total food production is contributed by Rainfed agriculture. Rainfed agriculture is complex, diverse and risk-prone. Most of the Small and Marginal Farmers (SMFs) live in Rainfed areas and continue cultivating their land only to retain their ownership (or user) rights. Farming without irrigation is a hard way of life (http://lnweb18.worldbank.org/ sar/sa.nsf). At the same time, they have an abundance of biomass and cattle-waste. Convergence and integration of traditional knowledge and practices/ systems are suggested for development of Rainfed agriculture.

The Agenda-21 of the Rio Earth Summit (1992), the UN Convention to Combat Desertification (CCD), the UN Convention on Biological Diversity (CBD), the UN Framework Convention on Climate Change (UNFCCC) and its Kyoto Protocol, and the Habitat Agenda adopted by the UN Conference on Human Settlements in 1996, directly or indirectly, suggested integrated planning and management of Water, Land, Minerals, and Biota resources (that land comprises), for sustainable development and use. Lester Brown (2002)[2] suggests an economy for the Earth – "Eco-economy" – that respects the principles of ecology, which can be integrated into ecosystems, in a way that will stabilize the relationship between the economy and the earth, enabling economic progress to continue. Sustainability of Land Resources (Figure-1) is therefore the key to food, water and livelihood security, biomass supply, healthy environment and social stability of a country.

The Nagpur Declaration (2000)[3] on "Natural Resources Planning and Management for Sustainable Development" suggested that both "river basins management" at the macro level and "watershed management" at the micro level should mutually complement each other, for integrated Water resources planning and management. A key element of the FARM Program[4] is the recognition of communities "indigenous

---

[2]Lester R. Brown (2002). "Eco-Economy: Building for the Earth," Orient Longman private Limited, India, 2002

[3]Nagpur Declaration (2000). denotes the Declaration of the Indian Geographers Congress, organised by the National Associations of Geographers India (NAGI), held at NBSS&LUP (Nagpur), during 1–3, January 2000.

[4]The Farmer-centered Agricultural Resource Management (FARM) Program is an initiative of eight Asian countries, viz. China, India, Indonesia, Nepal, Philippines, Sri Lanka, Thailand and Vietnam, and supported by UNDP and implemented by FAO, during 1990s.

knowledge, which when complemented by specialist "formal knowledge," promote participatory learning and research for achieving sustainable use and management of natural resources in agriculture and attainment of household food security through innovative approaches, in Rainfed areas.

For scientific utilization of natural resources base, it is considered that product of interaction of rain with land; in other words, watershed is the ideal geographical unit. Each watershed contains a complex mixture of: soil types, landscapes, climatic regimes, land use characteristics, and agricultural systems, and can be subdivided into agro-ecoregions (AER) having similar soil types, landscapes, climatic regimes, crop and animal productivity, and hydrologic characteristics.

The whole country has been divided into six river resources regions, 35 basins, 112 Catchments, 550 subCatchments and 3,237 Watersheds (AISLUS)[5], and then mini-watersheds and microwatersheds. Watershed development has become a trusted tool for the overall development of the village and people living within a watershed area.

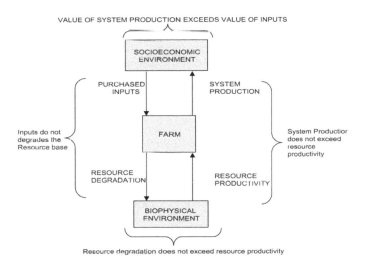

FIGURE 18.1   Sustainable Landuse System.

[5]AISLUS: All India Soil and Land Use Survey (AISLUS) Organization, now renamed as Soil and Land Use Survey of India (SLUSI), has brought out Watershed Atlas on 1:1,000,000 scale, which provides a uniform delineation and codification system of watersheds that could be followed by all concerned agencies dealing with watershed approach, on a common basis.

This is due the fact that the development is based on type of soil, depth of soil, vegetative cover, harvestable rain water in that area, watering that area, water budgeting, and treatment given to soils from the ridge to the valley. Basic Components of the Watershed Approach consists of following components:

i.   Community Development (Human Resource Development),
ii. Soil and Land Management,
iii. Water Management,
iv. Afforestation,
v.   Pasture/Fodder Development,
vi. Livestock Management,
vii. Rural Energy Management, and
viii. Farm and Non-Farm Value Addition activities.

In the words of eminent economist, Professor C.H. Hanumantha Rao, "Watershed development has been conceived basically as a strategy for protecting the livelihoods of the people inhabiting the fragile ecosystems experiencing soil erosion and moisture stress." This strategy has been designed with the objectives of public participation for conservation, upgradation and utilization of natural endowments (i.e., land, water, animal and human resources) in a harmonious and integrated manner, generation of massive employment in rural areas, improved standard of living of millions of poor farmers and landless laborers, restoration of ecological balance through scientific management of land and water, reduction of inequality between irrigated and Rainfed areas. The Ministry of Agriculture has established the National Rainfed Area Authority (NRAA) to meet this objective. The Planning Commission has published a Common Guidelines for undertaking Watershed Development Program (http://www.nraa.gov.in). Initially, this NRAA was established under the Ministry of Agriculture, and then shifted to the Ministry of Rural development, and now under the Planning Commission.

18.3   SUSTAINABLE DEVELOPMENT AND SUSTAINABLE LIFESTYLES

Energy security, food security and Climate Change are interlinked. Eliminating poverty, ensuring sustainable development and a cleaner energy future are among the foremost global objectives. India affirmed to promote the implementation of the UN Framework Convention on Climate Change (UNFCC) in accordance with the Bali Action Plan. The December 2009 saw the UN FCC's 15th Conference of Parties (COP15) at Copenhagen. There have been a considerable amount of awareness, education, research, networking, etc., on Climate Change (CC) in the country and elsewhere also.

Since 2000 onwards, as a technocrat interested in mainstreaming ICT for Grassroots level development and prosperity, I have been championing issues related to "sustainable Development." As the Secretary General of the Bhoovigyan Vikas

Foundation (BVF) and the Distinguished Fellow of the BVF, which is consortium of Earth Science and Social Science Professionals, working towards the common goal of "saving our planet," through various activities designed to achieve "sustainable development and sustainable life styles," conforming to the post-Agenda 21 scenario, I have organized the 1st International Conference on "Sustainable Lifestyles and Sustainable Development" on 22–23 April 2001 and the "Earth Day" was celebrated, in the Bal Bawan Public School – Mayur Vihar Phase II (Delhi) on April 21, 2001 with 5000 school children in Delhi, first time in India. The 2nd International Conference was targeted on "Sustainable Agriculture, Water Resources and Earth Care Polices" was held on 18–20 December 2002 in Delhi. The Broad subthemes of the 2nd Conference are discussed in the following subsections.

18.3.1 SUSTAINABLE AGRICULTURE

- Sustainability of Agriculture: Economic, social, institutional and technological dimensions
- Local issues, Micro-level planning and Ecofarming for sustainable agriculture.
- Sustainable agriculture in different agro-ecological systems-Irrigated, Arid, Rainfed, Hill and Mountain and coastal.
- Farming systems, Integrated Economic Circuits and provision of infrastructure (Capital intensive, Capital extensive and institutional) for sustainable agriculture.
- Lessons from Traditional Technologies.

Sustainable agriculture will provide the answer to our food security problem in the context of restructuring our economy to make it compatible with the Earth's ecosystem, so that economic progress can continue. Development of sustainable agriculture has many facets. Apart from the economic and social dimensions, it is linked to poverty alleviation, livelihood opportunities, utilization of common property resources, microlevel planning and development of local level institutions, including credit institutions and subsidy programs.

18.3.2 WATER RESOURCES

- Water resources: Projections on potentials and integrated use;
- Hydro-geology and Ground water use: Legal, moral and ethical issues in conflicting and complementary interests;
- Individual versus social and community rights in water resources use and management;
- In-situ and ex-situ water conservation including water harvesting: technologies and social issues;

- Water pollution, watershed management and Water use efficiency in irrigation projects;
- Drinking Water: Quantity, availability, financial resources, organizational/institutional setup, operations and maintenance (O&M).

In terrestrial ecosystems, Water constitutes the most important resource that limits plant growth and yield. There is broadly, a three-way demand on available fresh water for human use: (i) for agriculture, animal husbandry, fisheries, etc., to provide food, (ii) to meet the civic needs of people in rural and urban areas, and (iii) water for industrial enterprises. In these competing demands, water for agriculture, though most critical, tends to receive only a lower priority. Irrigation can vastly increase agricultural productivity and yields of most crops, notably cereals by 100 to 400%. However, poor water management poses many dangers. Water logging and salinization are major problems in irrigated agriculture. Besides, overexploitation of groundwater by excessive pumping decreases water table depth and in coastal areas intrusion of seawater occurs affecting water resources and water quality. Limited water resources are already a constraint to development in large parts of the country. Increasing productivity of water, as opposed to land productivity, requires a more diverse and creative mix of strategies, combining science, technology, information and knowledge-intensive management options.

What is required is an integrated policy for development of water resources and its utilization encouraging conjunctive water-use, limiting overexploitation of groundwater resources and regarding water as a common and national resource. At the micro level, we need to improve the efficiency of water use, adopt better watershed management, better water harvesting practices and reduce leakage, especially in the cities where water losses are alarming. Cities should not drain our villages.

18.3.3   EARTH CARE POLICIES

- S&T Development policies including legislative;
- S&T in Natural Resource Management Policy;
- Agencies and Institutions for Earth-Care Policies—Role of People's participation;
- Forest conservation and management including agro-forestry, Ecotourism and planned use of Wasteland & Degraded land;
- Earth-Care Policies for Fragile EcoSystems and resource-rich natural regions (e.g., North Eastern Region & Western Ghat in India).

The International Union for Conservation of Nature and Natural Resources (IUCN) has defined "Caring for the Earth" by emphasizing two requirements. One is to secure a widespread and deeply held commitment to the ethic for sustainable living and to translate its principles into practice. The other is to integrate conservation and development; conservation to keep our action within the earth's carrying capacity. Governments will have to adopt appropriate Earth Care Policies within the above framework to build a sustainable ecoeconomy. Now, we have a Central Ministry for Earth Sciences, after the Tsunami 2006 took place, in the Country.

18.3.4 FUSION OF TECHNOLOGIES FOR SUSTAINABLE AGRICULTURAL DEVELOPMENT

- Information and Communication Technology (ICT):ICT use and knowledge management, emerging learning technologies, decision support systems, global e-sustainability initiative and impact of digital economy;
- Biotechnology: Potential of biotechnology in food and agriculture, environmental impact of GM Crops and biosafety, and impact of Intellectual Property Rights;
- Ecotechnology: Environmental sustainability, blending traditional knowledge with frontier technologies and ecovillage network;
- Geomatics (Spatial Informatics) Technology: Geomatics and geo-related sciences, metadata standards, agricultural resource modeling, developments in sensor-controller technology for farm management and spatial knowledge modeling;
- Evaluation and Validation of Natives and Natural Products/Technologies.

18.4 EARTH CARE POLICY INSTITUTE (EPI) – THE NEED OF THE HOUR

During the first decade of the new Century, as we are still groping on how to reverse the environmental deterioration of the Earth, three imperatives prominently strike our attention viz. (i) control of population in our country; (ii) food security; and (iii) environmental security. In this context, it is apparent that through sustainable agriculture, water resources development and appropriate earth-care policies, we must seek to transform our economy shaped largely by market forces into one shaped by the Principles of Ecology.

Global projections indicate that the world population is expected to reach 8 billion by 2030 and accelerated rural-urban migration may lead to a situation of over half of the world's people living in cities or urban conglomerates. All these would increase the pressure to produce enough food to meet the needs of the additional 2 billion people while preserving and enhancing the natural resource base upon which the wellbeing of the present and future generations depends.

One of the greatest achievements of the human society has been the remarkable increase in land productivity from about 1.0 ton per ha in 1950 to over 2.8 tons per ha in the 1990s. This remarkable achievement, which has come about through the advent of the Green Revolution Technologies, has led many nations towards the path of self-sufficiency in food production. Evidence is now accumulating that this has been achieved at a great negative cost both economically and socially. At the biophysical level, the negative impacts have led to reduction in Total Factor Productivity (TFP), soil organic carbon depletion, increase in soil salinity, imbalances in water regimes, and pollution due to indiscriminate application of fertilizers and pesticides.

In its concept paper, the Bhoovigyan Vikas Foundation (BVF) (http://www.bhoovikas.org), with whom I have been closely associated, has suggested that India needs an EPI, in the similar lines of Earth Care Institute in Washington (USA), as a policy-oriented research institution and would adopt a 'Systems Engineering Approach' to the solution of environmental and earth care problems requiring an interdisciplinary and multilateral approach, involving both physical and social scientists and including studies on rural, urban and natural and human resource development problems. The BVF also suggested that this National Centre would draw upon the intellectual resources available in Earth Sciences and Social Sciences Departments of Universities and Research Institutes, specializing in Earth-related and Planning-oriented studies. For this purpose, the EPI will have three major groups or divisions with their distinct responsibilities, as given below:-

- A Division to undertake selected "Thematic and Policy-oriented Studies" with four departments:
  o Natural Resources Management, Sustainable Development/Consumption/Life Styles.
  o Geomatics applications and future studies.
  o Surveys and Applied Research.
  o Studies on Earth and Human Systems and Anthropogenic interventions.
  o Studies on climate change, ocean development and energy, are currently being undertaken by some specialized institutions.
- A "Spatial Database and Management Division" to continuously documenting all relevant data and information about the status of health and well-being of the Earth in its various ecosystems and will be assisting and facilitating all the research activities of the Centre.
- A "Outreach Division" to establish links with all academic and research institutions in the Country engaged in "Earth Science Studies and Research," and also institutional linkages with institutions abroad.

Such EPI is required in the Country. In addition this, I have been articulating in all my addresses in related forums to establish a "National University of Geography (NUG)" in the country. For sustainable development, we need education, research, development, training and extension in a coordinated manner. Each of these five pillars is orthogonal to each other now. They are required to be converged.

18.5 NATIONAL MISSION ON SUSTAINABLE AGRICULTURE (NMSA)—RELEVANT RECOMMENDATIONS

The Mission Document of National Mission on Sustainable Agriculture (NMSA) 2009, missions as given below:

  i. devise strategies to make Indian agriculture more resilient to climate change,
 ii. support the convergence and integration of traditional knowledge and practices systems,
iii. update and inventories of available region-wise technologies for farming system,
 iv. generate awareness through stakeholder consultations, training workshops and demonstration exercises for farming communities, for agro-climatic information sharing and dissemination,
  v. strengthen the National Seed Grid ensuring supply of seed across the country as per area specific requirements,
 vi. issue Soil Health Pass Book to each farmer with soil testing advisories within next five years, and
vii. undertake, in an organized manner, the production and marketing needs of Bio-fertilizers, liquid fertilizers and compost.

The National Action Plan on Climate Change (NAPCC) has dealt with "Information and Data Management" issue, in great details and also recommended database development projects for operationalization during 2012–2017. Its summary recommendations are as follows:

  (i) Extensive use of ICT for providing the required extension support to be promoted by linking all the Blocks by a Wide Area Network (WAN) and connectivity to be provided up to the village level through Common Service Centers (CSCs);
 (ii) Creation of detailed Soil Database to develop micro level agricultural land use plan and Space enabled spatial database for Village Resource Centers;
(iii) Development of a National Portal on agriculture statistics and soil resource and its spatial decision support systems.
 (iv) The land use statistics data, under nine fold classifications, to be extended to village/Panchayat level in pilot districts.
  (v) A prototype of ICT based information dissemination system to be developed for stakeholders for various agro-climate zones.
 (vi) Appropriate decision support system (DSS) for assessing risk and risk profiling at farm level, regional level as well as at national level including appropriate advisories for risk mitigation. Relevant information need to be converted to knowledge before it is transferred to the grass-root for wider impact and dissemination.
(vii) Relevant information need to be converted to knowledge before it is transferred to the grass-root for wider impact and dissemination.

(viii) Installation of about 20,000 Automatic weather Stations (AWSs) for collating weather data from Panchayat level, assuming that a weather station can be representative in about 5 k.m. radius;

(ix) Development of GIS and remote-sensing methodologies for detailed soil resource mapping and land use planning at the level of a watershed or a river basin;

18.5.1 ACTION POINTS OF SUB-MISSION ON RISK MANAGEMENT

Agricultural insurance is an important instrument for management of production risks. But the available options and insurance products are limited to mitigate risks that farmers are likely to be exposed to. The situation becomes worse because of the currently limited coverage and inadequate access to information to the farmers and other stakeholders. The designing and developing demand-driven agricultural insurance products requires evolution of efficient tools and refinement in methodology of accurate risk profiling. The novelty of linking quantitative risk to agricultural insurance products is that it takes care of variety of risk regimes protecting millions of small farmers against risks. The National Policy for Farmers 2007 provides for preparation of Drought Code, Flood Code and Good weather Code for drought prone, flood prone and other agricultural areas respectively, so that the required corrective/mitigation measures can be taken well in time.

For improving the risk management in agriculture sector, this Sub-Mission on Risk management recommended the following priority areas:-
i. Strengthening of current agricultural and weather insurance mechanisms;
ii. Development and validation of weather derivative models (by insurance providers ensuring their access to archival and current weather data);
iii. Creation of web-enabled, regional language based services for facilitation of weather-based insurance;
iv. Development of GIS and remote-sensing methodologies for detailed soil resource mapping and land use planning at the level of watershed or a river basin;
v. Mapping vulnerable ecoregions and pest and disease hotspots;
vi. Developing and implementing region-specific contingency plans based on vulnerability and risk scenarios.

Information and Communication Technology (ICT) requires to be used for collating information relating various weather, soil, water and pest related risks. Converting the information to knowledge and transforming into action required to prevent and manage the risk would be ensured through application of technology. Synergy in the existing efforts of various Ministries and Agencies to create a database with advisory for farmers to effectively manage the risks would be emphasized.

18.5.2 ACTION POINTS OF SUB-MISSION III: ACCESS TO INFORMATION

This Sub- Mission addresses problems related to Access to Information among different stakeholders in agriculture including farmers, development agent's researchers, and policy makers. The objective of this Sub-Mission is to fill the knowledge gap providing valuable information to all the stakeholders in time. The collaboration, partnerships and strategic alliances between, Department of Agriculture and Cooperation, research organizations and other related organizations can provide useful information.

It is well recognized that dissemination would generate recommendations for better decision-making. The knowledge warehouse for sustainable agriculture developed under this Sub-Mission will not only provide strong capabilities to the stakeholders, that is, farmers, researchers, decision makers to understand the impact of climatic changes but also help to build models for providing timely forecast of status of agricultural situation well in advance. The Recommended Actions Points arc as follows:

a. Fill the knowledge gap through creation of Data warehouse. In this regard, a comprehensive spatial database on various parameters related to land use, input use (seed, fertilizer, agricultural technology, agricultural credit), water usc, etc.; through convergence of data on soil, water and other natural resources available at various agencies at subdistrict level (district, block, Panchayat and village level) would be created/developed.

b. Develop Collaboration, partnerships and strategic alliances between, Department of Agriculture and Cooperation, research organizations and other related organizations.

c. Empower extension workers, farmers, researchers, decision makers to understand the impact of climatic changes and provide timely forecast of weather and agriculture situations well in advance.

d. Extensive use of ICT shall be provided up to the village level through Common Service shall be promoted by linking all the Blocks by a wide area network and connectivity Centers (CSCs).

e. Flow of information for pest surveillance and livestock diseases would be streamlined.

18.6 TERMS OF REFERENCE

1. Development of Regional Databases of Soil, Weather genotypes, land use pattern and water resources;

2. Collation and dissemination of block level data on agro-climatic variables, land use and socioeconomic features.

Under this Sub-Mission, the Department of Agriculture and Cooperation (DAC) proposed to collect data on agro climatic variables, land use and socio economic factors through NIC who in turn will use the database so developed as well the data warehouse developed by IASRI and other relevant sources such as IMD and National Centre for Medium Range weather forecasting (NCMRWF), etc., of the Ministry of Earth Sciences for developing user friendly software tools and necessary hardware, integration into web based framework and development of decision support systems (DSS). NIC has already developed the formats for compiling data on various aspects of agriculture and socioeconomic features which are available on net at www.disnic.nic.in. These formats will be modified in consultation with IASRI, IMD and NCMRWF in developing the block level data.

The activities under the present proposal envisage covering all aspects of agriculture and allied sectors including natural resources management to generate a number of deliverables. Agricultural Resources Information System (AgRIS) (http://agris.nic.in) of the Department of Agriculture & Cooperation (DAC) is envisaged to see the convergence of data on soil, water and other natural resources available at various agencies at subdistrict level (District, Block, Panchayat and village level). This project will further extend this convergence. Some of the important deliverables are to:

- Develop a comprehensive spatial database on various parameters related to land use, input use (seed, fertilizer, agricultural technology and agricultural credit), water use, etc.;
- Enable the extension workers and the farmers to access desired information for improving productivity;
- Capacity building, Farmers Training and Extension activities;
- Plan for weather contingencies;
- Access research and technology database;
- Reporting model and knowledge processing framework with every level of Decision Support System (DSS);

The Natural Resources Management (NRM) Division of ICAR, Natural Resources Management (NRM) Division of DAC and Agricultural Informatics Division of NIC were involved in the various deliberations of this Sub-Mission Group Meeting. NIC and IASRI are the main agencies in developing necessary Databases and DSSs. It is recommended that NIC will cooperate with the stakeholders (IASRI, NCMRWF, Ministry of Water Resources, Ministry of Rural Development, Ministry of Environment and Forests, I&B, Department of Space, State Departments, SAUs and ICAR Institutions, NGOs, etc.) through DAC.

The Planned layout for this Project is up to 2016–2017 and recommended outlay is about Rs. 1500.00 Crores. In order to proceed further, it has been suggested that NIC has to prepare a Detailed Project Report (DPR). NIC has requested DAC to release project amount to undertake preparation of DPR for this sub-Mission Component. It was proposed to cover 600 districts in a phased manner as follows:

Year	2009–2010	2010–2001	2011–2002	2012–2003	2013–2004	2014–2005	2015–2006	2016–2017	Total
No. of Districts to be covered	25	50	100	100	100	100	100	25	600

The Cost Estimates for developing software tools including hardware, integration into web based framework and development of decision support systems (Recurring and nonrecurring) for different activities proposed are given in Table 18.1.

TABLE 18.1 Cost Estimates for Recurring and Non-recurring Funds from 2009–2010 to 2016–2017

(Rs. In Crores)									
Funds	2009–2010	2010–2011	2011–2012	2012–2013	2013–2014	2014–2015	2015–2016	2016–2017	Total
Non-Recurring	50	100	200	200	200	200	200	50	1200
Recurring	Nil	12	20	28	38	52	65	85	300
Total	50	112	220	228	238	252	265	135	1500

Since the dissemination is proposed to be taken up by the strengthened extension system (up to village level) through National e-Governance Plan, no budgetary provision is made. NIC and IASRI are the main agencies in developing this information database. NIC through DAC will cooperate with following organization in this respect.

18.6.1 NATIONAL LEVEL

- Department of Agriculture and Cooperation of Ministry of agriculture – Lead Organization
- National Informatics Centre, Ministry of Communication and Information Technology
- Indian Agricultural Statistics Research Institute (IASRI)
- Indian Metrological Department, and National Centre for Medium Range Weather Forecasting (NCMRWF) of Ministry of Earth Sciences
- Ministry of Water Resources
- Ministry of Rural Development
- Ministry of Environment and Forests
- Ministry of Information and Broadcasting (Radio and TV)
- Department of Space

18.6.2 STATE LEVEL

- State Departments of Agriculture, and other line departments like Animal husbandry, Horticulture, Fisheries, Irrigation, Local Governments
- State Agricultural Universities and their research stations especially the Zonal Research Stations established at NARP Zones
- ICAR Institutes located in various states and their research stations including KVKs.
- Non-Governmental Organizations

At district and block level strong operational linkages are required to be established/ strengthened for all actors of ATMA.

18.6.3 ACTION POINTS OF THE SUB-GROUP-IV: PROMOTING DATA ACCESS

The National Action Plan on Climate Change (NAPCC) document has suggested "Promoting Data Access," as one of the broad themes of its National Mission on Strategic Knowledge for Climate Change. This document lists that there are several database, which are relevant for climate research and suggests that the respective agencies that are responsible for collecting and supplying such data may take action to digitize data, maintain database of global quality, and streamline the procedures governing access there to. It has further been suggested that the Ministry of Agriculture has to "expand and improve the existing database" on (a) Soil Profile (b) Area Under Cultivation, (c) Production and Yield, and (d) Cost of Cultivation.

The Department of Agriculture and Cooperation (DAC) has taken steps to digitize "the survey data as well as the relevant maps" by the Soil and Land Use Survey of India (SLUSI), in collaboration with National Informatics Centre (NIC). The National Informatics Centre (NIC) has already published its village level dataset designed, for its Projects for grassroots development: (a) DISNIC-PLAN Project: IT for Micro Level Planning (http://disnic.gov.in) and (b) Agricultural Resources Information System (AgRIS) Project (http://agris.nic.in) of Department of Agriculture and Cooperation pertains to Soil, Land, Groundwater and Environment parameters. Both these projects, DISNIC and AgRIS are under implementation in identified pilot districts.

However, the existing database on soil is inadequate to develop microlevel agricultural land use plan in the country, for which the needed scale of resolution is 1:4000/12,500. The detailed digital database on soil (physical, chemical and biological) is a prerequisite to address the various issues related to scientific Land Use Planning, soil reclamation; proper diagnosis of soils, judicious use of irrigation water and chemical fertilizers, nutrient deficiencies for maintenance of sound soil health and land productivity. The relevant soil parameters to be considered for such purposes are: soil type, elevation, type of land form, slope, geology (type of parent material), textural class, type of soil structure, soil water retentivity, soil pH, Electrical

Conductivity (EC), Organic Carbon, CaCO3, Fe %, Major oxides, available macro and micro nutrients, depth of water table, erosion class, drainage and runoff characteristics, land capability and irrigability, etc.

The Sub-Mission recommended its Plan of Action as: (a) Development of National Portal on Soil Resources including Detailed Soil Survey covering Block/Tehsil level data, (b) Soil Resource Mapping and Land Resource Planning, (c) Develop a National Portal on "Agricultural Statistics," (d) Outsourcing of Block/Tehsil wise LUS data pertaining to the last ten years and their digitization, (e) LUS data, under the nine-fold classification, at more disaggregated Levels (i.e., at the Village/Panchayat level) in pilot districts, and (f) Capacity Building through HRD for all Stakeholders of the System.

18.7   INSTITUTIONAL ARRANGEMENTS, COLLABORATING AGENCIES AND FINANCIAL OUTLAY

i.   Development of National Portal on Soil Resources (2009–2014) Institutional arrangement – NRM Divisions (ICAR & DAC), Directorate of Economics and Statistics (DES) – Ministry of Agriculture, and NIC Collaborating Agencies: SLUSI, NBSS&LUP, NRSA, NRAA, DOLR, NIC, State Land Use Boards, SASA, State Agricultural Universities, ICAR Institutes, Agricultural Colleges and Departments of Geography, etc. Financial Outlay: Rs. 50 crores

ii.   Development of "Detailed Soil Resource Mapping" and "Land Use Planning" (1:4000/12,500) (2009–2014) Institutional arrangement – NRM and DES of DAC and Public-Private-Panchayat-Partnership (PPPP) Model Collaborating Agencies – SLUSI, NBSS&LUP, NRSA, NRAA, DOLR, NIC, State Land Use Boards, SASA, State Agricultural Universities, ICAR Institutes, Agricultural Colleges, Departments of Geography and Common Services Centers (CSCs), etc.; Financial Outlay – Rs. 1000.00 crores

iii.   Development of National Portal, Improving Quality of Estimates, and Cost of Cultivation (Duration – 8 Years) Institutional Arrangements – Directorate of Economics and Statistics Collaborating Agencies – All Implementing Agencies/State Agricultural Statically Agencies (SASAs) M&E – Economic & Statistical Advisor Financial Outlay – Rs. 90.00 Crores

A joint proposal on (i) Development of National Portal on Soils and (ii) Soil Resource Mapping for Farm Planning in India, has been submitted to the Director General (ICAR) who is the Vice-Chairman of the NMSA.

18.8   RECOMMENDATIONS

The NMSA Report 2008 recommends undertaking Land Use Planning for Sustainable Agricultural and Rural development for effective production and productivity, and suggests creation of Natural Resources Management System using detailed soil resource mapping, and also Agricultural Resources Information System (AgRIS) at village level. To make Integrated Land Use Planning as a Decision Support System (DSS) tool, the Agricultural and Rural development Officers working at subdistrict levels and District levels need to undergo both Capacity Building and Capability Building Programs during the 12th Plan period. It is essential to look at the following recommendations:

a.   Capacity building of Block and District level Agricultural officers, Rural Development officers, Livestock officers and Fisheries officers, in the area of "Integrated Land Use Planning for Sustainable Agriculture and Rural Development." This amounts to training of about 25,000 officers during the next two years.

b.   Placement of one geographer in each Gram Panchayat for Natural Resource Accounting.

c.   Strong Emphasis on NRM in CDAP.

d.   Establishment of a National University of Geography (NUG), to be administered by NBSS & LUP and make their eight campuses as University campuses to bring in scientific based research in LUP for SARD.

e.   Involvement of 150 Departments of Geography in the NMSA project: Soil Resource Mapping and Agricultural Resources Information System (AgRIS) and SLUBs.

f.   Curriculum Changes are required in the Geography Discipline so that the Department of Space and ISRO can employ the Geography Students suitably in their organizations and also adopt these 150 Departments as their Laboratories for Remote Sensing Data Analysis for local needs.

g.   Introduction of Geography discipline at B.Tech Level (4-Year course) namely: Geographical sciences and Engineering."

h.   Setting up of a National Taskforce consisting of experts on Land Use Planning, Agricultural development and rural development, and Agricultural informatics and Rural informatics, to review these recommendations periodically and advise the State Governments through "Working Reports." The Aim should be to bring out at least 28 District level Reports, one per State, during the 12th Plan Period and to be utilized for district level planning and development during 2017–2022 or even during 2012–2017.

i.   Information Service on "Integrated Land Use Planning for Sustainable Agricultural and Rural Development," is to be introduced as SERVICE under the Agriculture Mission Mode project (AMMP), launched under the National e-Governance Program (NeGP), in collaboration with NBSS&LUP and State Land Use Boards.

18.9 CONCLUSION

The National Workshop on Integrated Land Use Planning for Sustainable Agricultural and Rural Development, organized by the Centre for Agrarian Studies and Disaster Mitigation of the National Institute of Rural Development and Panchayati Raj (NIRD & PR), has given an insight into the real problems of the Communities at grassroots level. Rural India has enormous budget outlay for its development and its "Per Village Capita" Development Fund can usher in "development" if Integrated Land Use Planning is adopted for sustainable development and inclusive growth. Will the Development Administration in districts utilize the services of Higher Secondary Schools, Colleges and Universities of the District, for its developmental planning and its development architecture? The District Annual plan outlay, which amounts to an average of Rs. 1200 Crores, thus, requires UTILISATION based on database based micro level planning. Development of Integrated Land Use Planning as a "Knowledge and Decision Support System" for Sustainable Agricultural and Rural Development through Academia and Government and Public participation is required at grassroots level, for achieving economic prosperity in Rural India.

KEYWORDS

- **Agricultural Resources Information System**
- **Agriculture Mission Mode Project**
- **Decision Support Systems**
- **Earth Care Policy Institute**
- **GIS Technology**
- **Informatics for Sustainable Agricultural Development**
- **Information Systems**
- **Knowledge and Decision Support System**
- **Land Use Planning**
- **Land Use Planning**
- **National e-Governance Program**
- **National Informatics Centre**
- **Natural Resources Management**
- **Principles of Ecology**
- **Rainfed Agriculture**
- **Remote Sensing Technology**
- **River Basins Management**
- **Watershed Management**

REFERENCE

M. Moni (2011). "Paradigm Shift in Geography: Challenges and Opportunities," a Keynote Address delivered at the National Symposium on "Paradigm Shift in Geography" organized by A.M. Khwaja Chair, Department of Geography, Jamia Milia Islamia, New Delhi, on 28th November, 2011.

CHAPTER 19

CLIMATE CHANGE AND RISK MANAGEMENT IN AGRICULTURE

V. U. M. RAO[1]

CONTENTS

[1]Central Research Institute for Dryland Agriculture, Santoshnagar, Hyderabad – 500 059

ABSTRACT

Global climate experienced accelerated climate changes during the last century primarily due to rapid industrialization and indiscriminate use of natural resources. As India's 60% population makes their earning from agriculture, country is more vulnerable to climatic changes. Rainfall behavior during the last 52 years showed spatial differences and increasing/decreasing trends in rainfall events. Accelerated warming was noted during last decade on a country basis. GHG emissions from agriculture sector are about 17%. Changes in climatic elements are found not only to influence agriculture but also allied sectors. In the present study, an attempt has been made to identify these impacts and strategies to minimize them are discussed. Insurance, both crop and weather, is discussed as one of the risk management strategy for Indian conditions.

19.1   INTRODUCTION

The debate on climate change was initiated as early as 1896 by noted Swedish chemist and Nobel Prize winner Svante Arrhenius, who proposed that a rise in average global surface air temperature by around 5–6 °C would be triggered by doubling of CO_2 in lower part of atmosphere. It is well accepted now that climate change will mean higher average temperatures, changing rainfall patterns and rising sea levels. These are also associated now with more and more intense extreme events such as droughts, floods and cyclones due to climate change. Although there is a lot of uncertainty about the location and magnitude of these changes, there is no doubt that they pose a major threat to agricultural production/systems. The Fourth Assessment Report (IPCC, 2007) stated that Earth's climate has changed in an unprecedented manner in the past 400,000 years, but accelerated greatly during the last century, due to rapid industrialization and indiscriminate destruction of natural environment. Agriculture is under tremendous pressure due to ill effects of climate change. Agriculture and forestry sectors contribute around 1/3rd of global warming potential and are also highly sensitive to climate change. Developing country like India is particularly vulnerable because its population depends directly on agriculture and natural eco systems for their livelihoods. The warming trend in India over the past 100 years (1901 to 2007) was observed to be 0.51 °C with accelerated warming of 0.21 °C per every 10 years since 1970 (Krishna Kumar, 2009). A more recent study indicated that annual minimum temperature is increasing @ 0.24 °C per 10 year on all India basis (Bapuji Rao et al., 2013). The projected impacts are likely to further aggravate yield fluctuations of many crops with impact on food security and prices. Climate change impacts are likely to vary in different parts of the country. Parts of Western Rajasthan, Southern Gujarat, Madhya Pradesh, Maharashtra, Northern Karnataka, Northern Andhra Pradesh, and Southern Bihar are likely to be more vulnerable in terms of extreme events (Mall et al., 2006a). Hence, there is a need to address the whole issue of climate change and its impacts on Indian agriculture in totality so as to mitigate the same through adaptive techniques against the global warming on war-footing.

Climate risk is a particular challenge for the hundreds of millions whose livelihoods depend on rainfed agriculture in marginal and high-risk environments. Rainfed agriculture is often characterized by high variability in production outcomes, that is, by production risk. Unlike most other entrepreneurs, farmers cannot predict with certainty on crop yields, due to external factors such as weather, pests, and diseases. Agricultural production can also be hindered by adverse events during harvesting that may result in production losses. For designing appropriate risk management policies, it is useful to understand strategies and mechanisms employed by producers to deal with risk, including the distinction between informal and formal risk management mechanisms and between ex-ante and ex-post strategies. The ex-ante or ex-post classification identifies the time in which the response to risk takes place: ex-ante responses take place before the potential harming event; ex-post responses take thereafter. Ex-ante informal strategies are characterized by diversification of income sources and choice of agricultural production strategy.

One innovative response to enable poverty reduction through better climate risk management is weather insurance. Weather insurance will continue to be the dominant insurance concept as the coming years are likely to experience more frequent extreme weather events like heavy rains, droughts, heat and cold waves, etc. Food security and weather risk management are inextricably linked: weather risk management or the lack of it determines the level of systematic risk in the food security system. At the farm level, weather based index insurance allows for more stable income streams and could thus be a way to protect people's livelihood and improve their access to finance. Weather based insurance is an upcoming strategy that has proven its worth in places such as India and it is important that it has to be given the attention it deserves to improve the food securities of communities especially the resource poor. Finally, climate change needs to be treated as a major economic and social risk to national economies, not just as a long-term environmental problem.

19.2 TRENDS IN KEY WEATHER PARAMETERS

Rainfall is the key variable influencing crop productivity in agricultural crops in general and rainfed crops in particular. Intermittent and prolonged droughts are a major cause of yield reduction in most crops. Long-term data for India indicates that rainfed areas witness 3–4 drought years in every 10-year period. Of these, 2–3 are in moderate and one may be of severe intensity. However, no definite trend is seen on the frequency of droughts as a result of climate change so far. For any R&D and policy initiatives, it is important to know the spatial distribution of drought events in the country. Analysis of number of rainy days based on the India Meteorological Department (IMD) grid data from 1957 to 2007 showed declining trends in

Chhattisgarh, Madhya Pradesh and Jammu Kashmir. In Chhattisgarh and Eastern Madhya Pradesh, both rainfall and number of rainy days are declining which is a cause of concern as this is a rainfed rice production system supporting large tribal population who have poor coping capabilities.

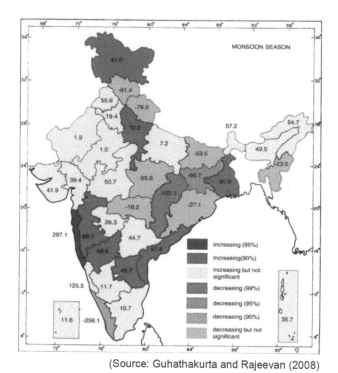

(Source: Guhathakurta and Rajeevan (2008)

FIGURE 19.1   Increase/Decrease in summer monsoon rainfall (mm) in India during last century (Source: Guhathakurta and Rajeevan (2008).

Guhathakurta and Rajeevan (2008) reported that all India summer monsoon (June to September) rainfall does not show any significant trend during the last century. However, three subdivisions *viz.*, Jharkhand, Chhattisgarh, Kerala show significant decreasing trend and eight subdivisions *viz.*, Gangetic West Bengal, West Uttar Pradesh, Jammu & Kashmir, Konkan & Goa, Madhya Maharashtra, Rayalaseema, Coastal Andhra Pradesh and North Interior Karnataka show significant increasing trends during last century (Fig. 19.1).

Temperature is another important variable influencing crop production particularly during *rabi* season. A general warming trend has been predicted for India but knowing temporal and spatial distribution of the trend is of equal importance. Lal (2001) reported that annual mean area-averaged surface warming over the Indian subcontinent is likely to range between 3.5 and 5.5 °C by 2080s (Table 19.1).

TABLE 19.1 Projected mean temperature changes over the Indian sub continent

Year	Season	Temperature change (°C)	
		Lowest	Highest
	Annual	1.00	1.41
2020s	Rabi	1.08	1.54
	Kharif	0.87	1.17
	Annual	2.23	2.87
2050s	Rabi	2.54	3.18
	Kharif	1.81	2.37
	Annual	3.53	5.55
2080s	Rabi	4.14	6.31
	Kharif	2.91	4.62

19.3 ROLE OF GREENHOUSE GASES IN CLIMATE CHANGE

The increasing levels of Greenhouse gases (GHG's) in the atmosphere have been attributed as one of the major driving force behind the rapid climate change phenomenon. The main GHG's contributing to this phenomenon is CO_2, CH_4 and N_2O. The contribution of India to the cumulative global CO_2 emissions from 1980 to 2003 is only 3.11%. Thus historically and at present India's share in the carbon stock in the atmosphere is relatively very small when compared to the population. India's carbon emissions per person are twentieth of those of the US and a tenth of most Western Europe and Japan.

Though the increase in the level of CO_2 is expected to produce some beneficial effects on crop production, it may soon be nullified by associated water and thermal stresses leading to overall deterioration of agro-climatic conditions for food production systems. At the global scale, the historical temperature-yield relationships indicate that warming from 1981 to 2002 is very likely to offset some of the yield gains from technological advances, rising CO_2 and other nonclimatic factors (Lobell and Field, 2007). The recent greenhouse gas (GHG) inventory released by Indian government has revealed that the Net GHG emissions from India in 2007 were 1727.71 million tons of CO_2 equivalent (eq) of which CO_2, CH_4, N_2O emissions were 1221.76, 20.56, 0.24 million tons, respectively. GHG emissions from agriculture sector alone constitute about 17% of the net CO_2 emissions.

19.4   EMISSION OF GREENHOUSE GASES FROM INDIAN AGRICULTURE

It is also important that role of agricultural activities in increasing the levels of GHG's is often overlooked. Assessment of GHG inventory that identifies and quantifies a country's primary anthropogenic sources and sinks of GHG emission is central to any climate change study. India being a party to the United Nation's Framework Convention on Climate Change (UNFCCC) needs to develop, periodically update, publish and make available to the Conference of Parties, a national inventory of anthropogenic emissions by sources and removals by sinks of all GHGs. Accordingly the inventory of GHG emission by Indian agriculture was developed for the base year 2000 (Table 19.2).

TABLE 19.2   Greenhouse Gas Inventory for Indian agriculture for the year 2000

Source	CH4 (Tg)	N2O (Gg)	CO2 eq. (Tg)
Ruminant	10.1	-	252.0
Rice cultivation	3.5	-	87.3
Manure management	0.1	0.1	2.5
Crop residue	0.2	4.0	4.9
Soil	-	132.3	39.4
Total	14.7	137.3	386.1

Tg = million ton, Gg = thousand ton.

19.5   IMPACT OF CLIMATE CHANGE ON AGRICULTURE

The impact of climate change on agriculture may accentuate at regional level creating more vulnerability in food security rather than global level as a whole. The potential impact will be shifts in sowing time and length of growing seasons, which may necessitate effective adjustment in sowing and harvesting windows, change in genetic traits of cultivars and sometimes total adjustment of cropping system itself. With warmer environment associated with erratic rainfall distribution, resulting higher rate of evaporation will enhance depletion of soil moisture and soil fertility. Hence for sustaining the crop productivity, efforts should be made to enhance the water and nutrient efficiencies by adopting resilient management practices. Apart from these tackling with frequent and more intense extreme events like heat and cold waves, droughts and floods may become norm of the day (IPCC, 2001). Such phenomena will impact agriculture considerably through their direct and indirect effects on crops, livestock, and incidences of pest-disease-weeds, increasing deterioration of soil health in totality and thereby threatening the food security like never before.

The output of the studies so far carried out by Aggarwal (2009) have indicated that a marginal 1 °C increase in atmospheric temperature along with increase in CO_2 concentration would cause very minimal reduction in wheat production of India if simple adaptation strategies like adjustment of planting date and varieties are adopted uniformly. But in absence of any adaptive mechanism the yield loss in wheat go up to 6 million tons. A further rise by 5 °C may cause loss of wheat production up to 27.5 million tons. Similarly, rice yields may decline by 6% for every one-degree increase in temperature (Saseendran et al., 2000).

In addition to direct effects on crops, climate change is likely to impact natural resources like soil and water. Increased rainfall intensity in some regions would cause more soil erosion leading to land degradation. Water requirement of crops is also likely to go up with projected warming; extreme events like floods, cyclones, heat wave and cold wave are likely to increase.

The availability of viable pollen, sufficient numbers of germinating pollen grains and successful growth of pollen tube to the ovule are of fundamental importance in grain formation. The network study on wheat and rice suggested that high temperature around flowering reduced fertility of pollen grains as well as pollen germination on stigma. These effects are more pronounced in Basmati rice as well as *Durum* wheat cultivars. A positive finding of the study was that the *Aestivum* wheat cultivars are more or less tolerant to such adverse affects. But differential impact of increasing temperature is observed with respect to grain quality of wheat where it is found that *Aestivum* wheat cultivars are more prone to reduced grain quality due to increasing temperature during the fruit setting stage than *Durum* cultivars. Field experiments using advanced 'Temperature gradient tunnels' with different dates of sowing to study impact of rising temperature on growth and development of different crops revealed that an increase of temperature from 1 to 4 °C reduced the grain yield of rice (0–49%), potato (5–40%), green gram (13–30%) and soybean (11–36%). However, one of the important pulse, chickpea, registered 7–25% increase in grain yield by an increase in temperature up to 3 °C, but was reduced by 13% with further 1°C rise in temperature.

A significant decrease in average productivity of apples in Kullu and Shimla districts of Himachal Pradesh has been reported which is attributed mainly to inadequate chilling required for fruit setting and development (Fig. 19.2). Reduction in cumulative chill units of coldest months might have caused shift of apple belt to higher elevations of Lahaul-Spiti and upper reaches of Kinnaur district of Himachal Pradesh. In general temperature below 7 °C for total 800–1400 h is taken as chilling requirement of apple; however temperature below 1 °C and above 18 °C is not desired for accumulating chill units.

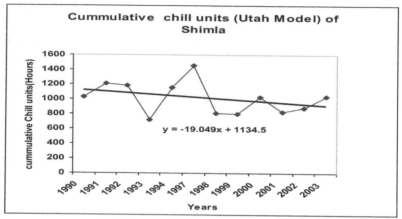

FIGURE 19.2   Linear trend of cumulative chill units for apple at Shimla.

The impact of rising temperature and CO_2 are also likely to change insect-pest dynamics. Dilution of critical nutrients in crop foliage may result in increased herbivore of insects. For example, Tobacco caterpillar (*Spodoptera litura*) consumed 39% more castor foliage under elevated CO_2 conditions than controlled treatments (Srinivasa Rao et al., 2009).

The nutrient loss from soil through high rate of mineralization and CO_2 emissions from soil could be accelerated as a result of increase in temperature. Low carbon soils of mainly dryland areas of India are likely to emit more CO_2 compared to high or medium carbon temperate region soils. Simulation of water balance using Global and Regional Climate Models revealed likely increase in annual as well as seasonal stream-flows of many Indian river basins pointing to the need for adoption of more effective runoff and soil loss control measures to sustain crop production across the country.

19.6   CLIMATE CHANGE IMPACT ON LIVESTOCK

India owns 57% of the world's buffalo population and 16% of the cattle population. It ranks first in the world in respect of cattle and buffalo population, third in sheep and second in goat population. This sector uses crop residues and agricultural by-products for animal feeding that are unfit for human consumption. Livestock sector has registered a compounded growth rate of more than 4.0% during last decade, in spite of the fact that a majority of the animals are reared under suboptimal conditions by marginal and small holders and milk productivity per animal is low. Increased heat stress associated with rising temperature may, however, cause distress to dairy animals and possibly impact milk production. A rise of 2–6 °C in temperature is expected to negatively impact growth, puberty and maturation of crossbred cattle and buffaloes. The low producing indigenous cattle are found to have high level of tolerance to these adverse impacts than high yielding crossbred cattle. Therefore, high producing crossbred cows and buffaloes will be affected more by climate change.

Research conducted at National Dairy Research Institute, Karnal revealed that livestock begin to suffer from mild heat stress when temperature-humidity index (THI) reaches higher than 72, moderate heat stress occurs at THI 80 and severe stress is observed after THI reaches 90. In India, huge variation in THI is observed throughout the year. In most of the agroclimatic zones of India, the average THI are more than 75. More than 85% places in India experiences moderate to high heat stress during April, May and June. THI ranges between 75 and 85 at 2.00 PM at most part of India. The THI increases and exceed 85, that is, severe stress levels at about 25% places in India during May and June. Even during morning THI level remains high during these months. On an average, THI exceed 75 at 75–80% places in India throughout the year. As can be seen from data, the congenial THI for production, that is, 70 is during Jan and Feb at most places in India and only about 10–15% places have optimum THI for livestock productivity, that is, during summer and hot humid season. Climate change scenario constructed for India revealed that temperature rise of about or more than 4 °C is likely to increase uncomfortable days (THI>80) from existing 40 days (10.9%) to 104 days (28.5%) for HadCM3 A2 scenario and 89 days for B2 scenario for time slices 2080–2100. The results further indicate that number of stress days with THI >80 will increase by 160%.

It is estimated that global warming is likely to result in a loss of 1.6 million tons in milk production by 2020 and 15 million tons by 2050. Based on THI, the estimated annual loss in milk production at the all-India level by 2020 is valued at Rs. 2662 crores at current prices. The economic losses were highest in Uttar Pradesh followed by Tamil Nadu, Rajasthan and West Bengal. Stressful THI with 20 h or more daily THI-hrs (THI > 84) for several weeks, affects animal responses. Under climate change scenario, increased number of stressful days with a change in maximum and minimum temperature and decline in availability of water will further impact animal productivity and health in Punjab, Rajasthan and Tamil Nadu (Upadhyay et al., 2009).

19.7   CLIMATE CHANGE IMPACT ON FISHERIES

A rise in temperature as small as 1°C could have important and rapid effect on the mortality of fish and their geographical distributions. Oil sardine fishery did not exist before 1976 in the northern latitudes and along the east coast of India as the resource was not available/and sea surface temperature (SST) were not congenial. With warming of sea surface, the oil sardine is able to find temperature to its preference especially in the northern latitudes and eastern longitudes, thereby extending the distributional boundaries and establishing fisheries in larger coastal areas (Vivekanandan et al., 2009a).

The dominant demersal fish, the threadfin breams have responded to increase in SST by shifting the spawning season off Chennai. During past 30 years period, the spawning activity of *Nemipterus japonicas* reduced in summer months and shifted towards cooler months. A similar trend was observed in *Nemipterus mesoprion* too. Analysis of historical data showed that the Indian mackerel is able to adapt to rise in sea surface temperature by extending distribution towards northern latitudes, and by descending to depths (Vivekanandan et al., 2009b).

Central Marine Fisheries Research Institute, Cochin has studied the vulnerability of 75 coastal fishing villages of Maharashtra, which are located within 100 m from the high tide line to sea level rise. Among the 75 coastal villages, five coastal districts (Thane, Mumbai, Raigad, Ratnagiri and Sindhudurg) of Maharashtra, it was found that 35 villages in Raigad and Ratnagiri districts would be affected due to rise in sea level by 0.3 m.

19.8   IMPACT OF CLIMATE CHANGE ON CROP WATER REQUIREMENTS AND WATER RESOURCES

Besides hastening crop maturity and reducing crop yields, increased temperatures will also increase crop water requirement. A study carried out by Rao et al. (2011) on the major crop growing districts in the country for four crops, *viz.*, wheat, maize, sorghum and pearl millet indicated a 2.2% increase in crop water requirement by 2020 and 5.5% by 2050 across all the crops/locations. The climate scenarios for 2020 and 2050 were obtained from HadCM3 model outputs using 1960–1990 as base line weather data (Table 19.3).

TABLE 19.3 Estimated crop water requirement (mm) of four crops in major growing districts of the country under projected climate change ccenario

District (State)	1990	2020	2050	% change over 1990 in	
				2020	2050
Wheat					
Sirsa (Haryana)	281.8	293.1	301.4	4.0	7.0
Ahmedabad (Gujarat)	523.0	536.8	551.0	2.6	5.4
Ahmednagar (Mah)	485.8	496.1	509.5	2.1	4.9
Ganganagar (Raj)	278.9	290.3	298.2	4.1	6.9
Hardoi (UP)	475.0	488.2	502.2	2.8	5.7
Kangra (HP)	367.7	380.7	391.2	3.5	6.4
Vidisha (MP)	437.1	446.9	460.4	2.3	5.3
Sangrur (Punjab)	391.1	405.4	416.3	3.7	6.4

TABLE 19.3 *(Continued)*

District (State)	1990	2020	2050	% change over 1990 in	
				2020	2050
Maize					
Udaipur (Raj)	388.8	392.4	400.9	0.9	3.1
Karimnagar (AP)	424.7	433.4	440.0	2.0	3.6
Jhabua (MP)	424.5	430.6	441.9	1.4	4.1
Begusarai (Bihar)	370.0	374.7	388.9	1.3	5.1
Bahraich (UP)	407.4	412.1	426.5	1.1	4.7
Godhra (Gujarat)	426.3	432.3	444.0	1.4	4.2
Khargaon (MP)	354.3	365.0	381.0	3.0	7.6
Aurangabad (Mah)	413.4	423.1	435.7	2.3	5.4
Sorghum					
Solapur (Maha)	348.8	373.2	399.9	7.0	14.7
Gulburga (Kar)	387.2	396.9	411.9	2.5	6.4
Khargaon (MP)	350.3	355.1	364.1	1.4	3.9
Mahabubnagar (Telangana)	383.4	393.7	409.1	2.7	6.7
Ajmer (Raj)	362.9	365.4	375.4	0.7	3.4
Coimbatore (TN)	378.8	387.8	396.9	2.4	4.8
Banda (UP)	326.7	332.5	347.0	1.8	6.2
Surat (Guj)	308.9	314.3	321.7	1.8	4.1
Pearl Millet					
Barmer (Raj)	337.8	338.6	347.4	0.2	2.8
Nashik (Maha)	284.2	289.9	296.9	2.0	4.5
Agra (UP)	277.5	279.7	289.7	0.8	4.4
Gulbarga (Kar)	325.2	333.5	344.0	2.6	5.8
Bhind (MP)	285.8	287.4	298.0	0.6	4.3
Villupuram (TN)	311.4	317.7	333.7	2.0	7.2

Next to agriculture and related to agricultural needs, is the water sector. It can be seen from the projections of future water requirements due to climate change that the current level of water availability is fast dwindling and may fail to meet the future water needs. The targets are fast increasing due to increased population demands and poverty eradication needs to be met by 2015, 2030 and 2050. Efficient irrigation and rain water management in rainfed areas will play a key role in minimizing the impacts and in improving agricultural productivity and also protect soil environment (Ramakrishna et al., 2007).

At present, available statistics on water demand shows that the agriculture sector is the largest consumer of water in India using 83% of the available water. The quantity of water used for agriculture has increased progressively through the years as more and more areas were brought under irrigation. Possible impact of climate change on water resources during the next century over India is furnished in Table 19.4. The enhanced surface warming over the Indian subcontinent by the end of the next century would result in an increase in premonsoonal and monsoonal rainfall and no substantial change in winter rainfall over the central plains. This would result in an increase in the monsoonal and annual run-off in the central plains, with no substantial change in winter run-off and increase in evaporation and soil wetness during the monsoon and on an annual basis (Mall et al., 2006b).

TABLE 19.4   Impact of climate change on water resources during the next century over India

Region/Location	Impact
Indian subcontinent	• Increase in monsoonal and annual run-off in the central plains • No substantial change in winter run-off • Increase in evaporation and soil wetness during monsoon and on an annual basis
Odisha and West Bengal	One meter sea-level rise would inundate 1700 km² of prime agricultural land
Indian coastline	One meter sea-level rise on the Indian coastline is likely to affect a total area of 5763 km² and put 7.1 million people at risk
All-India	Increases in potential evaporation across India
Central India	Basin located in a comparatively drier region is more sensitive to climatic changes
Kosi Basin	Decrease in discharge on the Kosi river; Decrease in run-off by 2–8%
Southern and Central India	Soil moisture increases marginally by 15–20% during monsoon months
Chenab River	Increase in discharge in the river
River basins of India	General reduction in the quantity of the available run-off, increase in Mahanadi and Brahmini basins
Damodar Basin	Decreased river flow
Rajasthan	Increase in evapotranspiration

19.9 ADAPTATION AND MITIGATION STRATEGIES

Successful adaptation to climate change requires long-term investments in strategic research and new policy initiatives that mainstream climate change adaptation into development planning. As a first step, we need to document all the indigenous practices rainfed farmers have been following over time for coping with climate change. Secondly, we need to quantify the adaptation and mitigation potential of the existing best bet practices for different crop and livestock production systems in different agro-ecological regions of the country. Thirdly, a long-term strategic research planning is required to evolve new tools and techniques including crop varieties and management practices that help in adaptation.

More recently during 2010, Indian Council of Agricultural Research (ICAR) has launched the National Initiative on Climate Resilient Agriculture (NICRA) as a comprehensive project covering strategic research, technology demonstration and capacity building. Targeted research on adaptation and mitigation is at nascent stage in India but based on knowledge already generated, some options for adaptation to climate variability induced effects like droughts, high temperatures, floods and seawater inundation can be suggested. These strategies fall into two broad categories *viz.,* (i) crop based and (ii) resource management based.

19.10 CROP-BASED STRATEGIES

Crop-based approaches include growing crops and varieties that fit into changed rainfall and seasons, development of varieties with changed duration that can overcome the transient effects of change, development of varieties for heat stress, drought and submergence tolerance; evolving varieties which respond positively in terms of growth and yield under high CO_2. In addition, varieties with high fertilizer and radiation use efficiency and also novel crops and varieties that can tolerate coastal salinity and sea water inundation are needed. Intercropping is a time-tested practice to cope with climate variability and climate change. If one crop fails due to floods or droughts, second crop gives some minimum assured returns for livelihood security. Germplasm of wild relatives and local land races could prove valuable source of climate resilient traits. We need to revisit the germplasm collected so far, which has tolerance to heat and cold stresses but not made use in the past due to low yield potential. Susheel Kumar (2006) provides a succinct account of breeding objectives under climate change in India.

19.11 STRATEGIES BASED ON RESOURCE MANAGEMENT

There are large number of options in soil, water and nutrient management technologies, which contribute to both adaptation and mitigation. Much of the research done in rainfed agriculture in India relates to conservation of soil and rainwater and drought proofing which is an ideal strategy for adaptation to climate change (Venkateswarlu et al., 2009). Important technologies include in situ moisture conservation, rainwater harvesting and recycling, efficient use of irrigation water, conservation agriculture, energy efficiency in agriculture and use of poor quality water. Watershed management is now considered an accepted strategy for development of rainfed agriculture. Watershed approach has many elements, which help both in adaptation and mitigation. For example, soil and water conservation works, farm ponds, check dams, etc., moderate the runoff and minimize floods during high intensity rainfall. The plantation of multipurpose trees in degraded lands helps in carbon sequestration. The crop and soil management practices can be tailored for both adaptation and mitigation at the landscape level.

19.12   CLIMATE RISK MANAGEMENT AND AGRICULTURE

Climate risk is a particular challenge for the hundreds of millions whose livelihoods depend on rainfed agriculture in marginal, high-risk environments. Climate risk is not a new phenomenon, and climate risk management in the broad sense has long been practiced. Farmers anticipate the rains, using indicators, and time their planting and inputs based on their best estimates; they install irrigation system if they can; and they reduce risk exposure by diversifying their livelihoods as far as possible (Decron, 1996; Ellis, 2000). Scientists have also sought ways to help manage the risk that climate presents. Agriculture research has developed crop varieties that are drought tolerant, for example, and soil management practices that increase soil moisture-holding capacity. Weather forecasts have been a major advance in helping people plan appropriately. Rainfed agriculture is often characterized by high variability of production outcomes, that is, by production risk. Unlike most other entrepreneurs, farmers cannot predict with certainty the amount of output their production process will yield, due to external factors such as weather, pests, and diseases. Farmers can also be hindered by adverse events during harvesting or collecting that may result in production losses. In discussing how to design appropriate risk management policies, it is useful to understand strategies and mechanisms employed by producers to deal with risk, including the distinction between informal and formal risk management mechanisms and between ex-ante and ex-post strategies. The ex-ante or ex-post classification identifies the time in which the response to risk takes place: ex-ante responses take place before the potential harming event; ex-post responses take thereafter. Ex-ante informal strategies are characterized by diversification of income sources and choice of agricultural production strategy. One strategy producers employ is risk avoidance. Extreme poverty, in many cases, makes producers very risk-averse, pushing them to avoid high-risk activities, even though the income gains to be generated might be far greater than those got through less risky choices. This inability to accept and manage risk respectively reflected in the inability to accumulate and retain wealth is sometimes referred to as the "the poverty trap" (World Bank, 2001).

According to Clarkson et al., (2000), there are six requirements that must be met if rainfed farmers are to manage risks related to climate extremes, variability and change. These include:

- awareness that weather and climate extremes, variability and change will impact on farm operations;
- understanding of weather and climate processes, including the causes of climate variability and change;
- historical knowledge of weather extremes and climate variability for the location of the farm operations;
- analytical tools to describe the weather extremes and climate variability;
- forecasting tools or access to early warning and forecast conditions, to give advance notice of likely extreme events and seasonal anomalies;
- ability to apply the warnings and forecasts in decision making.

19.13 AGRICULTURE INSURANCE

Agriculture insurance, including livestock insurance has been practiced in the country for over 25 years. Prior to 2002–03, General Insurance Corporation of India (GIC) was implementing National Agricultural Insurance Scheme (NAIS). Recognizing the necessity for a focused development of crop insurance program in the country and an exclusive organization to carry it forward, Government created an exclusive organization – Agriculture Insurance Company of India Limited (AIC) on 20th December 2002 (www.aicofindia.org). AIC commenced business from 1st April 2003. AIC introduced rainfall insurance known as 'Varsha Bima' during the 2004 Southwest monsoon period. Varsha Bima (Varsha Bima covers anticipated shortfall in crop yield on account of deficit rainfall. Varsha Bima is voluntary for all classes of cultivators who stand to lose financially upon adverse incidence of rainfall can take insurance under the scheme. Initially Varsha Bima is meant for cultivators for whom NAIS is voluntary) provided for five different options suiting varied requirements of farming community. These are:

1. seasonal rainfall insurance based on aggregate rainfall from June to September;
2. sowing failure insurance based on rainfall between 15th June and 15th August;
3. rainfall distribution insurance with weights assigned to different weeks between June and September;
4. agronomic index constructed on the basis of water requirement of crops at different pheno-phases; and
5. catastrophe option, covering extremely adverse deviations of 50% and above in rainfall during the season.

Varsha Bima has been piloted in 20 raingauge areas spread over Andhra Pradesh, Karnataka, Rajasthan and Uttar Pradesh. During 2005, Varsha Bima was fine-tuned and extended to 120 locations in 10 States during *Kharif* 2005, and further to 150 locations in 15 States during *Kharif* 2006. AIC also introduced weather insurance pilots on wheat insurance, mango insurance, and coffee insurance during 2005–2006, and is looking ahead for expansion. Thanks to the sustained efforts and launch of innovative insurance products in the past 6 to 7 years, the number of farmers and the cropped area covered under crop insurance has seen spectacular growth. For the year 2009–2010, as many as 27 million farmers growing crops on over 38 million hectares were insured under various crop insurance programs of AIC. The numbers seem very impressive, yet a great majority of farmers who actually need insurance protection are still outside its purview.

19.14   CROP INSURANCE

The idea of crop insurance emerged in India during the early part of the twentieth century. Yet it was not operated in a big way till recent years. J.S. Chakravarti proposed a rain insurance scheme for the Mysore State and for India as a whole with view to insuring farmers against drought during 1920s. Crop insurance received more attention after India's independence in 1947. The subject as discussed in 1947 by the Central Legislature and the then Minister of Food and Agriculture, Dr. Rajendra Prasad gave an assurance that the government would examine the possibility of crop and cattle insurance. In October 1965, the Government of India decided to introduce a Crop Insurance Bill and a Model Scheme of Crop Insurance in order to enable the States to introduce, if they so desire, crop insurance. In 1970 the draft Bill and the Model Scheme were referred to an Expert Committee headed by Dr. Dharm Narain. Different experiments on crop insurance on a limited, ad-hoc and scattered scale started from 1972–1973. The first crop insurance program was on H-4 cotton in Gujarat. All such programs, however, resulted in considerable financial losses. The program(s) covered 3110 farmers for a premium of Rs. 4,54,000 and paid claims of Rs. 3.79 millions. It was realized that programs based on the individual farm approach would not be viable in the country. Obviously, "individual farm approach" would reflect crop losses on realistic basis and hence, most desirable, but, in Indian conditions, implementing a crop insurance scheme at "individual farm unit level" is beset with problems, such as:

1. non availability of past record of land surveys, ownership, tenancy and yields at individual farm level;
2. large number of farm holdings (nearly 116 millions) with small farm holding size (country average of 1.41 hectares);
3. remoteness of villages and inaccessibility of farm-holdings;
4. large variety of crops, varied agro-climatic conditions and package of practices;

5. simultaneous harvesting of crops all over the country;
6. effort required in collection of small amount of premium from large number of farmers;
7. prohibitive cost of manpower and infrastructure.

19.15 WEATHER INSURANCE

Weather insurance is an insurance coverage against the vagaries of weather. Many agrarian economies owe their strength to favorable weather parameters, such as rainfall, temperature, relative humidity, etc. Around 65% of Indian agriculture is heavily dependent on rainfall, and, therefore, is extremely weather sensitive. Many agricultural inputs such as soil, seeds, fertilizer, management practices, etc., contribute to productivity. However, weather, particularly rainfall has overriding importance over all other inputs. The reason is simple – without proper rainfall, the contributory value of all the other inputs diminishes substantially. An analysis of Indian Crop Insurance Program between 1985 and 2003 reveals that rainfall accounted for nearly 95% claims – 85% because of deficit rainfall and 10% because of excess rainfall (AIC, 2006). Reducing vulnerability to weather in developing countries may very well be the most critical challenge facing development in the new millennium. One of the most obvious applications of weather risk management products, weather insurance or weather derivatives. Weather impacts on many aspects of the agricultural supply and demand chain. From the supply side, weather risk management can help to control both production risk and quality risk. Weather events like warmer than normal winter or a cooler than normal summer can impact all sorts of companies like utilities, food and agricultural groups and even retailers. The basic idea of weather insurance is to estimate the percentage deviation in crop output due to adverse deviations in weather conditions. There are statistical techniques to work out the relationships between crop output and weather parameters.

19.16 WEATHER-INDEX

Index-based weather risk insurance contracts in agriculture have emerged as an alternative to traditional crop insurance. As the weather index is objectively measured and is the same for all farmers, the problem of adverse selection is minimized. Weather-indexed insurance can help farmers protect their overall income rather than the yield of a specific crop, improve their risk profile enhancing access to bank credit, and hence reduce overall vulnerability. Some of pilot schemes and delivery models operated in India are:
1. ICICI Lombard pilot scheme for groundnut in Andhra Pradesh;
2. KBS pilot scheme for soya farmers in Ujjain;
3. Rajasthan government insurance for orange crop;
4. IFFCO-TOKIO monsoon insurance;

5. AIC Varsha Bima Yojana (rainfall insurance scheme);
6. AIC Sookha Suraksha Kavach (drought protection shield);
7. AIC coffee rainfall index and area yield insurance;
8. ICICI Lombard loan portfolio insurance.

KEYWORDS

- **Agriculture Insurance**
- **Agriculture Insurance Company of India Limited**
- **Climate Change**
- **Crop Insurance**
- **General Insurance Corporation of India**
- **Greenhouse Gases**
- **Indian Agriculture**
- **Individual Farm Approach**
- **National Agricultural Insurance Scheme**
- **National Initiative on Climate Resilient Agriculture**
- **Sea Surface Temperature**
- **Temperature-Humidity Index**
- **The Poverty Trap**
- **United Nation's Framework Convention on Climate Change**
- **Weather Index**
- **Weather Insurance**

REFERENCES

Agarwal, P. K. (2009). Global Climate change and Indian agriculture; Case studies from ICAR network project. Indian Council of Agricultural Research. 148p.

AIC. (2006). Crop Insurance in India. Agriculture Insurance Company of India Limited (AIC), New Delhi, p.10.

Bapuji Rao, B., Santhibhushan Chowdary, P., Sandeep, V.M., Rao, V. U. M. and Venkateswarlu, B. (2013). Rising minimum temperature trends over India in recent decades: Implications for agricultural production. Gloabl and Planteary Change. 117, 1-8.

Clarkson, N. M., Abawi, G. Y., Graham, L. B., Chiew, F. H. S., James, R. A., Clewett, J. F., George, D. A., Berry, D. (2000). Seasonal stream flow forecasts to improve management of water resources: Major issues and future directions in Australia. Proceedings of the 26th National and third International Hydrology and Water Resource Symposium. The Institution of Engineers. Perth. pp. 653–658.

CRIDA. (2009). Annual Report 2009. Central Research Institute for Dryland Agriculture, Indian Council of Agricultural Research, New Delhi.

Dercon S. (1996). Risk, crop choice, and savings: Evidence from Tanzania. Economic Development and Cultural Change 44: 485–513.

Ellis F. (2000). Rural livelihoods and Diversity in Developing Countries. Oxford University Press, Oxford.

Guhathakurta, P., Rajeevan, M. (2008). Trends in rainfall pattern over India. International Journal of Climatology. 28, 1453–1469.

IPCC. (2001). Climate Change 2001: The Scientific Basis. Contribution of Working Group – I to the Third Assessment Report of the IPCC. [Houghton, J. T., Ding, Y., Griggs, D. J., Noguer, M., van der Linden, P. J., Dai, X., Maskell, K. Johnson, C. A. (Eds)]. Cambridge University Press, United Kingdom and New York, USA, 94p.

IPCC. (2007). Intergovernmental Panel on Climate Change 2007:Synthesis Report.p.23.

Krishna Kumar. (2009). Impact of climate change on India's monsoon climate and development of high-resolution climate change scenarios for India. Presented at MoEF, New Delhi on October 14, 2009 (http: moef.nic.in).

Lal, R. (2001). Future climate change: Implications for India Summer monsoon and its variability. Current Science. 81, 1205–1207.

Lobell, D. B., Field, C. B. (2007). Global scale climate-crop yield relationships and the impacts of recent past. Environmental Research Letters. 2:7.

Mall, R. K., Gupta, A., Singh, R., Singh, R. S., Rathore, L. S. (2006a). Water resources and climate change: An Indian perspective. Current Science. 90 (12), 1610–1626.

Mall, R., Singh, R., Gupta, A., Srinivasan, G., Rathore, L. (2006b). Impact of Climate Change on Indian Agriculture: A Review. Climatic Change. 78: 445–478.

Ramakrishna, Y. S., Rao, G. G. S. N., Rao, V. U. M., Rao, A. V. M. S., Rao, K. V. (2007). Water Management- Water Use in Rainfed Regions of India. In: Managing Weather and Climate Risks in Agriculture (Eds. Sivakumar, M. V. K. Motha, R. P.). Springer. pp.245–262.

Rao, V. U. M., Rao, A. V. M. S., Rao, G. G. S. N., Satyanarayana, T., Manikandan, N., Venkateshwarlu, B. (2011). Impact of climate change on crop water requirements and adaptation strategies, p311 to 319, (In) S. D. Attri, L. S. Rathore, M. V. K. Sivakumar and S. K. Dash. (Eds) Challenges and opportunities in agrometeorology, Springer, New York.

Saseendran, A. S. K., Singh, K. K., Rathore, L. S., Singh, S. V., Sinha, S. K. (2000). Effects of climate change on rice production in the tropical humid climate of Kerala, India. Climate Change. 44, 495–514.

Srinivasa Rao, Ch., Ravindra Chary, G., Venkateswarlu, B., Vittal, K. P. R., Prasad, J. V. N. S., Sumanta Kundu, Singh, S. R., Gajanan, G. N., Sharma, R. A., Deshpande, A. N., Patel, J. J., Balaguravaiah, G. (2009). Carbon sequestration strategies under rainfed production systems of India. Central Research Institute for Dryland Agriculture, Hyderabad. 102p.

Susheel Kumar. (2006). Climate change and crop breeding objectives in the 20 first century. Current Science. 90:1053–1054.

Upadhyay R. C., Ashutosh, Raina, V. S., Singh, S. V. (2009). Impact of climate change on reproductive functions of cattle and buffaloes. In: Global Climate Change and Indian Agriculture: Case Studies from the ICAR Network Project (Ed. P. K. Aggarwal). ICAR Publication.pp.107–110.

Venkateswarlu, B., Arun K. Shankar, Gogoi, A. K. (2009). Climate change adaptation and mitigation in Indian agriculture. Proceedings of National Seminar on Climate Change Adaptation Strategies in Agriculture and Allied Sectors. Kerala Agricultural University, Thrissur, December 2008, pp.109–121.

Vivekanandan, E., Hussain Ali, M., Rajagopalan, M. (2009b). Impact of rise in seawater tempera-ture on the pawning of threadfin beams. In: Global Climate Change and Indian Agriculture: Case Studies from the ICAR Network Project (Ed. P. K. Aggarwal). ICAR Publication. pp.93–96.

Vivekanandan, E., Rajagopalan, M., Pillai, N. G. K. (2009a). Recent trends in sea surface tem-perature and its impact on oil sardine. In: Global Climate Change and Indian Agriculture: Case Studies from the ICAR Network Project (Ed. P. K. Aggarwal). ICAR Publication. 89–92.

World Bank. (2001). World Development Report 2000/2001: Attacking Poverty. Washington.

CHAPTER 20

LAND USE SYNERGY AND LAND USE PLANNING: AN APPROACH FOR SUSTAINABLE RURAL DEVELOPMENT

ARUN CHATURVEDI[1], NITIN PATIL[2], and TRILOK HAJARE[3]

CONTENTS

[1]Head, Division of LUP, NBSS and LUP, Amravati Road, Shankarnagar P.O. Nagpur–440033.
[2]Senior Scientist, Division of LUP, NBSS and LUP, Amravati Road, Shankarnagar P.O. Nagpur–440033.
[3]Principal Scientist, Division of LUP, NBSS and LUP, Amravati Road, Shankarnagar P.O. Nagpur–440033.

ABSTRACT

Traditional sectoral land use based approaches have proven to be ineffective in addressing sustainable development goals. It is argued that if further environmental and social impacts are to be avoided without loss of agricultural productivity, there must be a synergy between different land uses. The current approach disregards the multiple and complex roles of land in development of the country and subsequently the need for cooperation across the many sectors involved. We argue for synergy between disparate agencies working towards a common goal of rural development and synergistic landscape approach for natural resources based development. It needs to be recognized that sectoral approach has several shortcomings and modern world demands services like livelihood, recreation, water supply, environment from the same parcel of land. Sustainable landscape-planning approach can contribute to solve several inherent shortcomings of the sectoral planning paradigm, including lack of integration of related agencies (such as forest) and the need for transdisciplinarity. It is emphasized that rural development can be sustainable if multifunctionality of agriculture is recognized and steps are taken in time to change the planning process failing which the conflicts are likely to escalate between competing land uses, a potentially damaging situation for any society aspiring to prosper.

20.1   INTRODUCTION

Land use till now has been a single focus domain, for instance agriculture land use is treated as a single function domain by policy makers and administrators. Forest land use is an exclusive territory wherein others uses are subjugated. However, there is now growing realization that a single function land use is a narrow perspective and there need to be recognition that any piece of land can provide different services like livelihood, recreation, environmental services, biodiversity conservation and so on. The multi functional land use in synergistic mode is perhaps the best option for managing such conflicting interests. The increasing pressure on the physically finite land resources has once again assigned importance to land use issues. As the per capita land availability continues to decline and the demand keeps on rising due to rising affluence, land use conflicts shall also keep on increasing, creating a divisive society. Urban encroachment on agriculture lands, agriculture encroaching forests, forests being a hindrance to irrigation/development and other infrastructure projects, including exploitation of mineral resources are realities we have to live and cannot be wished away. A country like India with more than 120 billion population has to face the toughest challenge compared any other country in the world. Therefore land use synergy has far greater significance to our environment, food security and general well being as a nation. In recent years, there has been a substantial shift towards recognizing that any area of land can provide many different environmental, recreational and health services at the same time and so therefore be considered 'multifunctional.'

The concept of land use synergy assumes importance especially when there is intense pressure on land and is thus being applied in urban land use in form of mixed zone systems. Interaction of land uses has probably existed since the first settlements had people who performed different types of work. Older towns and cities had all different types of uses within walking distances since walking was the principal mode of transportation. When suburbanization started to occur in the late 1800s, there began to be separations of different land use types. By the mid twentieth century, zoning and single-use areas had become the normal way to develop (http://www.greenexercise.org/ Multifunctional%20_Land_Use.html).

Later urban developers found the mixed – or multiuse developments appealing because such developments offered a way to capture several types of development in one project that was larger than any single project they might create in the same place. Moreover, the interaction and sharing of facilities had the potential to reduce long-term development costs and increase profitability. Development of urban agriculture in west is one of such development. The key to success is synergy between the land uses.

20.2   LAND USE SYNERGY

Land use synergy can be defined as multipurpose utilization or convergence of applications in designated land use areas, that is, rural functions in urban areas as is happening through farming/agriculture activities in the western countries. Similarly a lot of agriculture activity is taking place in protected forests and agro-forestry in rural areas. With merely 3.5% of Haryana's area under forests, the state has become self-sufficient in small wood, fuelwood and industrial timber by establishing large-scale plantations on farmlands. Trees in agro-ecosystems have increased the extent of area under forest and tree cover to 6.63%. These plantations sustain about 670 wood-based veneer, plywood and board, manufacturing units, one large paper mill and about 4300 sawmills that depend on agro-forestry produce. Bihar on the other hand, has 13 districts with well-developed irrigation facilities with an average of 56% of cultivated area under irrigation, well above national average (33%). All these districts are devoid of forests. Unlike Haryana, they do not have any accessible market to buy their wood needs, no local produce to meet such requirements. Plantations on the farm bunds alone could have helped to fulfill the requirement to great extent. In other parts of the country, forest canopy sometimes exceeds surrounding rural forests areas (as in Nagpur). In fact city like Nagpur has better green cover (Chaturvedi et al., 2013). In central India, permanent pasture and other grazing land increased from 6.68 million ha in 1950–1951 to 10.59 million ha in 2001–2002 constituting 2.3% and 3.5% of total reporting area and has decreased to 10.15 million ha. in 2009–2010. Permanent pasture and grazing land has shown a decline after

1980–81. This has led to shrinkage of grazing lands and other Common Property Resources (CPR's) resulting in pressure on the adjoining forestlands wherever available. Thus land use has consistently disregarded sustainability. Development while promoting environmental protection is a tough challenge especially in developing countries. Modern land use and changes in land use are mainly driven by economic interests though social interest could also be an influencing factor. The concept of mixed use that creates mutually supporting and complementary synergies among land uses has distinct advantages over single use zoning. A variety of land uses can be blended together in one area. The key is the effect of synergy must result in multiplier effect of combined/mixed land use. For instance, benefits of mixed use of land for agriculture and forest should be more than the sum of individual benefit. A mixed and integrated rather than a sectoral approach is an effective way of preventing or resolving the conflicting land uses. It optimizes planning process by involving all kinds of stakeholders as well.

Interestingly in the Indian perception, backwardness is associated with rural areas. While in reality, all backward areas are rural but all rural areas are not backward. In the same context, forest areas are also areas of high incidence of poverty and they do not count either in urban or rural areas. All nonurban areas, therefore, are either forest areas or agriculture based rural economies. Thus rural development in India account for either agriculture or forest disregarding the fact that these two land uses coexist in almost half the geographical area of the country and play significant role in livelihood of the inhabitants.

A lot of changes have taken place in recent times, which have provided new opportunities for serious debate on the livelihood potential of forests and their multifunctionality in utilization. It is important that if natural resources like land and water are to be optimally used for rural poverty alleviation, then the multifunctional role of each land use is appreciated and not treated as conflicting domains. For instance land use policy has been shown to be important in shaping bioenergy production (Rokityanskiya, et al. (2007), Gillingham et al. (2008), Wise, et al. (2009), Mellilo, et al. (2009)). Area dedicated to energy crops is likely to increase in many countries. Given the dependence on imported fossil fuel and likely higher returns, it can be presumed that energy plantation will be a conflicting land use if sectoral policies are continued. Environmental concerns and mitigation measures to combat climate change has also ushered a new thinking that emphasizes credence to reduction of Greenhouse gas emission (GHG) or GHG services to decelerate land degradation and desertification. The carbon credits generated by the forest and fringe dwellers can fetch them funds much required for development. Carbon sequestration through various means such as afforestation, reforestation, silvipasture, etc., or wastelands development, agro-forestry and other carbon conservation activities, and soil carbon conservation will soon be traded.

In fact, even small green technologies like sustainable use of bio-fuels, reduced use of chemical fertilizers, livestock management with ecofriendly technologies, methane based projects in rural area are likely to be recognized as tradable activities. There are many synergies in wasteland development/restoration of degraded lands and mitigation of GHG. Agriculture would thus become multifunctional. These changes are inevitable with only rider of time variation depending on the willingness of the governments to take suitable steps for developing synergies between forests and agriculture. Sudden delinking of tribal from forest particularly rights to NTFP could be disastrous. The process therefore has to be gradual with step-by-step approach to introduce alternatives and concomitant green technologies. Such changes require completely different outlook and thinking beyond the routine development paradigms.

The newly introduced legislations like 'The Scheduled Tribes and other Traditional Forest Dwellers (Recognition of Forest Rights Act) 2006, the National Policy for Farmers 2007 and other new anthropogenic policy initiatives have also necessitated a debate on the need for synergies between forest and agriculture. While the former has institutionalized and legalized agriculture practices in forest areas the latter has now included Pastoralists/ Herders/Migratory Graziers/Tribal Families/ Persons engaged in shifting cultivation and in the collection, use and sale of minor and nontimber forest produce also in the list of farmers. Thus whether it is climate change challenges or tribal development issues, the land use synergies have a very critical role in shaping the future policies.

The former President of India Dr. A.P.J. Abdul Kalam had mentioned that "One of the solutions for rural and urban disparities in the country could be reduced with PURA," which basically meant taking some of the urban amenities to the rural poor including drudgery reduction. This can be achieved by development in continuum through land use synergies. Isolated and stand alone land use systems will always create conflicts in resource consumptions while synergetic systems will play a complimentary role.

20.3 LAND USE PLANNING

Land use planning is the process of evaluating land and alternative patterns of land use and other physical, social and economic conditions for the purposes of selecting and adopting those kinds of land use and course of action best suited to achieve specified objectives. Land use planning may be at national, regional, state, district, watershed or village and also at farm level. The techniques and even the strategy of land use planning can be very different at national level, at village level and at any level in between. The process involves the participation of the land users and several other stakeholders. It entails systematically evaluating land and alternative pattern of land use, choosing that use which meets specified goals, and the drawing up of policies and programs for the use of the land.

Thus Land Use Plan can be targeted for increasing the agriculture production, for conservation of soils for improving the productivity of water in a command area/

watershed, or for ensuring livelihood, generating employment and last but not the least for ensuring food security. The methodology and drivers for each of these is different but the ultimate aim is to evaluate the land and present land use and select those combinations that best achieve the desired goals. According to David Dent (1981), any guide lines cannot be a prescription for land use planning, it should be flexible involving local or national procedures that can be modified by the people on the spot. The land use planning should encompass wide range of situations in which it can be applied. It has always been a matter of debate if Land Use Planning is possible in any democratic state where the absolute owner has absolute rights over the property. He may use the property as he likes. Thankfully, there are certain compulsions, which restrict his rights to use the property as he likes. The restrictions are imposed under various levels, mostly in common interest of public in large. The issues of human development & ecological preservation are not exclusive but mutually inclusive in today's world. It is ironic that land utilization which "manifests the man-environment relationship at any given time and place" is not given the same importance. Among the success stories for land use planning are the coffee, tea and rubber plantations in the colonial periods. These were in controlled administrative setup and directed to achieve optimum economic exploitation. Land use planning in the name of Master Plans in Urban areas is another example of optimum utilization of land to fulfill the aspirations of the local populace.

Land has various uses and these are guided by the human desire to meet certain aspirations, may it be economic, ecological, social or bound by external factors like international commitments. Some choices of land use for the present level of 'development' where land is used for the following purpose:

- maintenance of forest cover up to a certain specific level for ensuring ecological stability;
- production of cereals and other food crops for ensuring food security;
- production of fodder for the live stock that ensures supply of other food stuffs necessary for human beings;
- maintenance of water bodies for irrigation and supply of water for drinking and other purposes;
- cultivation of commercial crops for use as input by other production activities;
- provision of shelter to the population both in rural and urban area;
- extraction of minerals; and
- establishing industrial enterprises.

The list of purposes for which land is needed will keep on increasing as the demands of the society develop. An example is the additional requirement of land for biofuels, which is part of our international commitment. Some states (Tamilnadu and Chhattisgarh) started offering subsidy for energy plantation on marginal lands. However, reports indicate that private plantation companies are embezzling the subsidies by purchasing the land from poor farmers first, availing subsidy and after some time selling it to real estate developers at much higher prices. The central Government is targeting 20 million ha energy plantation with subsidized seedling and land preparation. The figure has been decided on the basis of perceived wastelands in the country. Such standalone land use planning perspectives, however, overlook minor facts like ownerships, which is essential to be established for subsidy disbursal. Wastelands in India are mostly common lands or under the control of Forest departments. In such a scenario, synergy in planning of land use systems especially through integrated landscape planning approach is the ideal solution.

20.4   INTEGRATED LANDSCAPE PLANNING

A coalition of agricultural and environmental organizations has released a new report recommending an "integrated or whole landscape" approach. It's an approach that brings together diverse and competing groups to discuss common needs. "This whole approach came out of the realities of competing interests – that in agricultural landscapes it's not just the farmers that want the water and it's not just one set of farmers that want the water. That maybe the crop farmers are competing with the livestock farmers for water and they're competing with the local towns for water. So a lot of these alliances that we're talking about started off as conflict situations or at least a lot of tempers being raised on a regular basis, " (Dr. Sara Scherer, an agricultural and natural resource economist and founder, president and CEO of Eco-Agriculture Partner (http://www.voanews.com/content/decapua-africa-landscapes-19jun12/1212884.html). Even landscape planning and land use zoning approach can coexist depending on the acceptability among the stakeholders. Landscape planning addresses issues larger than any single use and considers social, economical and environmental concerns in a better-organized manner. It is also expected that landscape planning will improve coordination between multiple management/development agencies and improve collaboration with the users and stakeholders. Prioritized use of limited resources can also improve overall planning efficiency. There could still be zones within the landscape or even protected forest, wildlife sanctuaries, etc. It is apparent that administrative set up in the country will have to incur many changes and reforms in institutional set up, development of new linkages, social acceptability will result in differential movement that perhaps will heighten the conflicts initially. However, adjustments will have to be made to achieve millennium goals and more importantly feed the burgeoning population.

Hawkins and Selman (2002) suggested that the basis of this (landscape) planning approach is to map at various scales those elements in the landscape that are inherently stable or unstable, and to determine from these maps a network of landscape elements to act as 'biocenters' and 'biocorridors' (Bucek et al., 1996). The existing network can then be analyzed to identify where landscape creation or rehabilitation is necessary to fill strategic gaps (Fig. 20.1). The basic concept involves retaining existing ecological infrastructure, and then creating 'more of the same' landscape elements in deficient areas. The work by Hawkins and Selaman has an urban planning context, and approach as in Fig. 20.1 may not have full relevance to India because of the token participation of people in land use decisions. However, as the current events indicate community may no longer accept decisions from the top regarding land use or at least land acquisition.

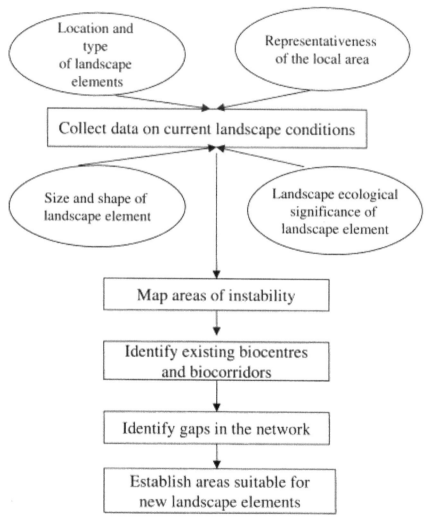

FIGURE 20.1   The Landscape Stabilization Model Adopted from Hawkins and Selman (2002).

According to Ferreira and Leitão (http://edepot.wur.nl/137347), the sustainable landscape-planning framework's main purpose is to provide a common planning framework applicable to all spatial planning activities. It has landscape ecology as its core scientific foundation. It constitutes an integration platform for a wide array of scientific knowledge that shares a spatial dimension and a landscape approach, including natural and social sciences and humanities. They have argued in favor of transdisciplinary approach that provides links between science, planning, management and stakeholders (Botequilha Leitão, 2001). The sustainable landscape-planning framework (Botequilha Leitão and Ahern, 2002) refers to sustainability under a more conceptual perspective by considering (i) the integrity of ecological systems as the basis for all human development; (ii) advocating for indispensable integration of the human component; and (iii) promoting the integration of social sciences and humanities. While these arguments sound very logical, in Indian context, it is very difficult to implement as disparate elements pull development process in different directions and yet stepwise advancements are contemporary requirement.

20.5 CONCLUSION

In a land scarce country like India agriculture can no longer be treated as an isolated, single function land use. A paradigm shift is necessary to overcome the impending challenges of continued competing demand for land for different uses without sacrificing other goals such as environment services, recreation, etc. Landscape approach is possibly the best way to deal with the natural resources based rural development.

KEYWORDS

- **Common Property Resources**
- **Greenhouse gas emission**
- **Integrated Landscape Planning**
- **Land Use Planning**
- **Land Use Synergy**

REFERENCES

Botequilha Leitão, A. Ahern, J. (2002). Applying landscape ecological concepts and metrics in sustainable landscape planning. Landscape and Urban Planning, 59 (2), 65–93.

Botequilha Leitão, A., 2001. Sustainable land planning: towards a planning framework method: exploring the role of landscape statistics as an operational planning tool. Universidade Tecnica de Lisboa, Lisbon. Ph.D. Dissertation.

Bucek, A., Lacina, J., Michal, I., (1996). An Ecological Network in the Czech Republic. Veronica special 11th issue. Dent, D., Young, A. 1981. Soil survey and land evaluation. London, Allen and Unwin. 278 pp.

Chaturvedi Arun, Kamble Rahul, Patil N. G., Chaturvedi Alka (2013). City – Forest Relationship in Nagpur, One of the Greenest City of India. Urban Forestry and Urban Greening 12(1), 79–87. http://dx.doi.org/10.1016/j.ufug.2012.09.003.

Gillingham KT, Smith SJ and Sands RD. 2008. Impact of bioenergy crops in a carbon dioxide constrained world: an application of the MiniCAM energy-agriculture and land use model. Mitigation and Adaptation Strategies for Global Change 13, pp. 675–701.

Hawkins V., Selman P., (2002). Landscape scale planning: exploring alternative land use scenarios Landscape and Urban Planning Volume 60, Issue 4, 30 August 2002, Pages 211–224.

http://www.greenexercise.org/Multifunctional%20_Land_Use.html Last accessed on July 2013.

http://www.voanews.com/content/decapua-africa-landscapes-19jun12/1212884.html. Last accessed on July 2013.

Melillo, J. M., Reilly, J. M., Kicklighter, D. W., Gurgel, A. C., Cronin, T. W., Paltsev, Felzer, B. S., Wang, X., Sokolov, A. P., Schlosser, C. A. (2009). Indirect Emissions from Biofuels: How Important? Science 326(5958), 1397–1399.

Rokityanskiya, Dmitry, Pablo C. Benítezb, Florian Kraxnera, Ian McCalluma, Michael Obersteinera, Ewald Rametsteinera, Yoshiki Yamagatac. (2007). Geographically explicit global modeling of land-use change, carbon sequestration, and biomass supply, Technological Forecasting and Social Change, 74(7), 1057–1082.

Wise, M. A., Calvin, K. V., Thomson, A. M., Clarke, L. E., Bond-Lamberty, B., Sands, R. D., Smith, S. J., Janetos, A. J., Edmonds, J. A. (2009). The implications of limiting CO_2 concentrations for land use and energy. Science 324, 1183.

CHAPTER 21

LAND USE POLICY FOR SUSTAINABLE DEVELOPMENT—SPECIAL REFERENCE TO UTTAR PRADESH

MRIDULA SINGH[1]

CONTENTS

[1]Additional Director, State Land Use Board, Government of Uttar Pradesh.

ABSTRACT

Any development is sustainable only when an optimal and efficient plan to use natural resources is firmly in place. Today the research focus is on the dynamics of land use. Land represents an important resource for the economic life of a majority of people in the world. The way people handle and use land resource is decisive for their social and economic well-being as well as for the sustained quality of land resources. This implies that policy discussion and development planning need to be based on a sound understanding of these dynamics. Conflict over land use is inevitable. With high population growth, endemic poverty and weak existing institutional capacity for land management a scientific and strategically important land use policy is must for sustainable development.

21.1   INTRODUCTION

Land utilization is a complicated issue and to some extent represents the man-environment relationship. In addition to the role of land and water, in terms of quantity as well as quality, the other important aspects in any land utilization are the management practices, the human dimension which influences and ultimately decides the success or efficiency of land use. Any development is sustainable only when an optimal and efficient plan to use natural resources is firmly in place. In this context it is widely perceived that any State can only prosper and have sustainable growth pattern if it uses its natural resources properly and judiciously. Otherwise even with the most productive soils and abundant water, economic poverty and backwardness will prevail. In the backdrop of critical issues, that is, climate change, food security and self reliance, use and reliability of information technology tools, available and proposed legal framework, mobilization, capacity building and knowledge enhancement related to land use planning it is a challenge and uphill task for policy makers and planners to recommend an acceptable and implementable land use policy framework.

It has been very appropriately said:

"In Asia, poverty has mainly been rural phenomenon and nearly three-fourth of the poor live in rural areas, with large majority of them dependent on natural resources for employment and income. South Asia, which had a poverty incidence of 43%, contributed about 40% of the world's poor. Development of natural resources thus offers a potentially enormous means of poverty reduction. Nowhere does this situation manifest itself explicitly as in the state of Bihar and Uttar Pradesh."

Uttar Pradesh, with a population of almost 200 million and **decennial growth rate** of 20%, is India's most populous state and also one of its poorest. UP's economy is dominated by agriculture, which represents a 40% share of the state's GDP and 65% of its employment. Uttar Pradesh has a major share in India's total food grain production and contributes 20.15% of total food grains from 7.36% of total area, with 16.5% population to support. In respect of net sown and gross cropped area (GCA), States share is 12.1% and 13.1%, respectively.

Out of the total geographical area of 328.02 m ha of India and 29.44 m ha of U.P., 17.02 m ha and 13.58 m ha area suffers from various kinds of degradation respectively. The per capita availability of agricultural land in India has decreased from 0.46 ha in 1951 to 0.26 ha in 1981 with likelihood of further decreasing to 0.15 ha. The situation of U.P. in this regard is further poor. Due to increasing population pressure, number of persons per hectare of net cropped area also increased from about 3 in 1951 to 5 in 1990, which is estimated to further increase to about 8 persons by the year 2025. Thus, the amount of land available for food supply and other human needs will continue to decline in per capita terms. (Rathore and Pal, 2000).

21.2 RESOURCE SCENARIO

The Indo-Gangetic Alluvial plains are among the most extensive fluvial plains in the world and extend in several states of northern, central and eastern parts of India. These are however not uniform in nature and the soils developed in alluvium exist under variety of climate and landforms, ranging from arid to subhumid and elevations of 150 m above MSL in Bengal Plain and 300 m above MSL in Punjab Plain with local variations in geomorphology deciding the soil type (Velayutham et al., 2002). The major part of the state (nearly two-third) is covered by what are popularly known as Alluvial Soils, with the level of Alkalinity being a crucial factor in their utilization. The Central-Western part of UP is also severally affected by problems of salinity and sodicity, the penalty being paid for irrational use of water for indiscriminate irrigation, though these were the areas which flourished during the Green revolution.

The dominant soils of Uttar Pradesh referred to as Benchmark Soils are varied and there are problems associated with their utilization. A growing concern in the state is the declining productivity of food grains, especially rice and wheat, which is mainly due to water-induced land degradation such as sodification, groundwater depletion, and loss of soil fertility. Sodification, especially in poorly managed irrigated areas, has left an estimated 1.25 million hectares of land completely barren. A further 1.25 million hectares of low-yielding salt-affected lands cover about 10% of the net cultivated area of Uttar Pradesh.

In spite of many central and state governments reclamation programs, results have been uneven due to limited institutional coordination, inadequate understanding of the total technology package, and a top-down approach that gave little encouragement to beneficiary participation. Efforts included the allotment of cultivable salt-affected land to landless laborers, who were largely unable to bring the lands back to full production.

> In Uttar Pradesh, high population pressure, extreme dependence on agriculture, declining soil fertility, climate change and environmental degradation have resulted in backwardness and poverty. Efforts are on in 2013–2014 to increase agricultural production by improving resource efficiency, input use efficiency, ensuring timely availability of quality inputs. The objective is to develop and popularize appropriate ecofriendly farming systems to improve soil health and farm income.

It is essential that regionally differentiated strategies are pursued taking into account the agronomic, agro ecological and socioeconomic conditions to realize full growth potential of every region. Improving soil health is another priority of the State Government.

21.3   LAND USE POLICY

Environmental conservation has been an integral part of the Indian ethos and is reflected in the Indian Constitution, adopted in 1950. Articles 48A and 51G of the Directive Principles of State Policy enjoin upon the State to protect and improve the environment and safeguard the forests and wildlife. The Constitution of India also enables the Centre and the States to enact laws to carry out the duties of preservation, afforestation and conservation of natural resources. Article 39(b) and (c) lays down the duty of the State and the Centre to develop natural resources for common good. Article 40, on the other hand, calls for organization of village as units of self-government. The Directive Principles of State Policy, though not enforceable by any Court, are nevertheless fundamental to the governance of the country. Thus, a favorable atmosphere for empowering grassroots communities and for assisting them to take initiatives in the areas of environmental management, including combating desertification, already exists.

Land and Water are subjects within the purview of the States. However, 'Forest' was a State subject earlier, and was brought to the Concurrent list in 1976. The subject 'Environment' is not under any List but is covered under the Directive Principles of State Policy and Fundamental Duties enshrined in the Constitution 'to protect and improve the environment.'

The list of central policies which have a bearing on land use, include:

• National Water Policy, 1987
• National Land Use Policy Outlines, 1988
• National Forest Policy (NFP) of 1988
• Policy Statement of Abatement of Pollution, 1992
• National Livestock Policy Perspective, 1996
• National Agricultural Policy, 2000
• National Population Policy, 2000
• National Land Reforms Policy, 2013

- National Policy and Macro-level Strategy and Action Plan on Biodiversity, 2000

In addition, there are legislative frameworks, which have to be conformed to by any state while planning a land use policy. These include:

- Forest (Conservation) Act, 1980
- Environment (Protection) Act, 1986
- Water (Prevention & Control of Pollution) Act, 1974 as amended in 1988
- Wildlife (Protection), 1972 as amended in 1988
- Constitutional Amendments (73rd and 74th Amendments) of 1992
- Municipality Act, 1992 (74th Amendment Act, 1992)

21.4 NATIONAL LAND USE POLICY

The "National Land Use Policy guideline and action points" were prepared by the Government of India, Ministry of Agriculture after intensive deliberations In the said policy, framing of suitable legislation and its sincere enforcement were stressed by imposing penalties, of violation thereof. The said policy guidelines were placed before the 'National Land Use and Waste Land Development Council' under the chairmanship of Prime Minister and its first meeting was held on 6th February, 1986. The Council agreed to the adoption of policy and circulated the same throughout the country for adoption after suitable considerations at the state level. Of the 19 points, some of the most important ones are:

1. Land Use Boards at the State level should be revitalized.
2. Land Use Policy must be evolved by all users of land within Government jointly and must be enforced on the basis of both legislation for enforcing land use as well as their promotional and preserving methods.
3. Urban Policy must be restructured so as to ensure that highly productive land is not taken away. Town planning should also provide for green belts.
4. A national campaign should be launched for educating the farmers and Government Departments about the need to conform to an integrated land use policy.
5. Land and soil surveys should be completed and inventory of land resources should be prepared in each State so that resources allocation is based on a reliable database.
6. Heavy penalties should be imposed against those who interfere with land resources and its productivity. It must be recognized that environmental protection cannot succeed unless this is done.
7. The problems of water logging, salinity and alkalinity must be brought under control by the use of appropriate technologies and by the adoption of proper water management practices.

8. Land Use Planning should be integrated with rural employment programs in such a manner that loans and subsidies are given only for those productive activities, which represent efficient land use.
9. Rights of tribal and poorer sections on common land should be protected through legal and administrative structures.

The policy was finally adopted in 1988 but has not really been able to move things. The Policy has been circulated to all concerned for adoption and implementation through enactment of suitable legislation. The policy, however, did not make the desired impact, mainly due to the fragmented handling of different components of agriculture like land and soil.

Any policy or regulating act prepared for a region should conform to the prevalent situation of the area. The basic purpose of any such piece of legislation is to protect the natural resources from being overexploited at any given time and also to improve the living conditions of the inhabitants. Land related policies and regulating acts, however, are perceived differently as they restrict and regulate activities, which may be beneficial to individuals at the moment, but are detrimental to the cause of society in the long run. It is here that the long-term benefits of sustainability of resources needs to be emphasized.

The framework for the Land Use Policy to be developed by each state is already available in the Guidelines for Land use Policy (1988) of the Government of India. The states are now required to fit in the particular scenario that exists in the region. These could be a policy for land use in the Usaror Ravine areas of U.P. Once such policies are framed, suitable legislations can be enacted within their realm. Problem-based decision-making is likely to be more successful than other approaches. Each state thus needs to look into:

• Scenario of land vis-a-vis land use: status, trends, pressures, driving forces;
• Consequences of land degradation and its effect on the society and global consequences;
• Effective and strong action for sustainable future policy.

The issue of human development and ecological conservation needs to be balanced for the overall good of the society. While legislations need more justification the decision makers need a concise, credible, legitimate, scientific assessment of the evidence, which is a difficult task when it comes to the issues of absolutes and relatives.

Guiding principles of Land Use Policy should be its formulation in accordance with the geographical, climatic and soil conditions (Agro-ecological regions) in the State. The policy of the State should strive to achieve self-reliance, self-generating economy, sustainability and conservation, development and management of eco system, particularly land resources viz. soil–water–plant–animal subsystems, betterment of socioeconomic conditions of the people, increase productivity and enhance production.

21.5   LAND USE POLICY FOR UTTAR PRADESH

As any land use plan has to be location specific, while drafting land use policy for the state following should be taken into account.

21.6   *OBJECTIVES OF LAND USE POLICY*

- Resource use efficiency to meet the growing consumption needs.
- To restore productivity of degraded lands.
- Suitable institutional mechanism for scientific management, conservation & development of land resources.
- Plan and resource linkages for land related program.
- Expansion of effective forest covers to restore ecological balance.
- Conjunctive use of surface and ground water resources.
- Translate policies through plan efforts into action.
- Greater awareness through education, training, extension programs.

21.7   ISSUES FOR CONSIDERATION

- Strengthening and capacity development of State land use boards.
- Achievement of desirable land use pattern through sectoral approach/plan linkages.
- Formulation of economical viable projects for each sector, that is, forests, agriculture, horticulture to translate land care into people's movement.
- Application of modern science and technology to have systematic and reliable database.
- Preparation of Land Use Atlas.
- Generation of strong political/administrative will.
- Computerized and updated Land Records.
- Strict laws/rules for land use conversion.
- Maintaining and improving soil fertility and unit area land productivity.
- Updated Surveys of land resources – climate, water, soils, landforms, forests, and rangelands to improve efficiency of investment.
- Regular Training/orientation programs for land use practitioners.
- Publicity of success stories in case of soil conservation and better land management.
- Effective reclamation of degraded lands.
- Measures to check further degradation of land.
- Effective watershed management and reduction in disparities and regional imbalances prevalent in the State through policy interventions.
- Diversification of land use.

- Preventive measures on adverse effects from industrial waste and effluents on good forest and cultivable land.
- Development of rural agro-based industries.
- Conservation of prime agricultural land through law.

KEYWORDS

- **Agro-Ecological Regions**
- **Alluvial Soils**
- **Indo-Gangetic Alluvial plains**
- **Land Use Policy**

REFERENCES

Srivastava S. K., Bandopadhyaya, S., Meena Rani, H. C., Hedge, V. S., Jayaraman, V. (2002). "Incidence of Poverty, Natural Resources Degradation and Economic Policies and Interventions: A Study Based on Wasteland Mapping. IAPRS & SIS, Vol. 34, Part 7. Resources and Environmental Monitoring, NRSA, Hyderabad, India.

Velayutham, M., Pal, D. K., Bhattacharya, et al. (2002). "Soils of the Indo Gangetic Plains India—The Historical Perspective."

CHAPTER 22

WORKSHOP RECOMMENDATIONS

M. V. RAO, V. SURESH BABU, K. SUMAN CHANDRA, and
G. RAVINDRA CHARY

CONTENTS

22.1   INTRODUCTION

Land is one of the prime resources for economic development of majority of the people in India. Problems with regard to land continue to attract special attention from policy-makers, development professionals, land use practitioners and academics. India today is facing a critical situation in relation to land use planning and land use management. Even though the foodgrain production recorded almost a fourfold increase in the postindependence era, food security is likely to emerge as a major challenge owing to near stagnancy in net sown area and rapid land and environmental degradation. Unless special efforts are made towards preservation of the natural resources of the country and its long-term sustainable use is planned, food security and self-reliance cannot be assured and enhanced livelihood security cannot be ensured. The National Commission on Agriculture (1976) has emphasized on detailed soil and land survey and scientific land use planning to achieve desired results. The National Commission on farmers (2007) has also recommended to prevent diversion of prime agricultural land and forest for nonagricultural purposes and to establish a National Land Use Advisory Service, which would have the capacity to link land use decisions with ecological meteorological and marketing factors on a location and season specific basis. Optimal Land Use is therefore a highly dynamic process, which implies that several related policies need to be based on a sound understanding of the potential of land and other related resources.

The advanced technologies like Geomatics and GIS is expected to guide the farmers by providing the information on land utility based on slope, soil type, depth, rainfall, moisture holding capacity, crop growth period, etc., at 1:10,000 scale which has been demonstrated in a few blocks in Tamil Nadu successfully. The land use pattern can be based on the land capability governed classification by adopting GIS tools. A new land use policy is required to govern the acquisition of land for industrial use (SEZs), infrastructure development and urbanization. Finalization of this legislation with an appropriate balance between the need to protect the interest of current land owners and those dependent on it for livelihood will lead to achieve the broader objectives of inclusive growth and development. Accordingly, the nonerable land classified under land capability class – VII and VIII may be allocated for industrial use or SEZs. Land management choices based on multilayers such as slope, soil type, rainfall pattern, moisture holding capacities; crop growth period will definitely minimize the risks and crop failure disasters and ensure sustainability.

The Centre for Agrarian Studies and Disaster Mitigation, National Institute of Rural Development and Panchayati Raj, Hyderabad conducted a National Workshop on "Integrated Land Use Planning for Sustainable Agriculture and Rural Development" during June 18–20, 2012. About 56 delegates participated in the workshop representing Administrators, Academics, Land Use Practitioners, Scientific Community from ICAR institutions and reputed NGOs. Twenty-one papers were presented.

The delegates all discussed in four working groups to deliberate on the major themes such as: (i) Status of land resources, (ii) Agriculture land use planning, (iii) SEZ land use and rural development, and (iv) Land use for rural development (non-agriculture).

The working groups have formulated the recommendations and action to be taken by the concerned organizations/departments, which was finalized under the chairmanship of Dr. M.V. Rao, IAS, Director General, NIRD & PR. The recommendations of the workshop have been broadly categorized into 10 subheads, which are mentioned hereunder:

1. Information needs on Natural resources for development of Land Use Plan (LUP);
2. Analyzing Scientific and remote sensing data (multiple layer data analysis by scientific and Research institutions at various levels);
3. Policy implication of scientific databases on optimum land use plan;
4. Legal framework;
5. Awareness and mobilizing farmer community;
6. Network of training institutions;
7. Capacity building of officials; elected representatives and farming communities;
8. Supply of scientific databases for each Gram Panchayat/Village;
9. Facilitating Land use planning and comprehensive plans preparation;
10. Convergence of Rural Development, Agriculture, Land and Water schemes and their monitoring.

22.2 RECOMMENDATIONS AND ACTION POINTS

22.2.1 INFORMATION NEEDS ON NATURAL RESOURCES FOR DEVELOPMENT OF LAND USE PLAN (LUP)

1. The existing soil maps (Land Resource Maps) available at 1:250,000 scale at State levels (and very few districts with 1:50,000 scale) are not adequate for developing detailed farm level planning and land use at farm level. Therefore, there is an urgent need for launching a country wide detailed soil and land resources survey for developing land resource maps of 1:10,000 scale for generating perspective land use plan at State and District levels and also pragmatic farm level land use planning and management at village/watershed level and creating a national portal on Natural Resources (Soil, Land, Water, Climate and land use capability) of India.
   Action: This mission mode program may be implemented in consortium approach with full participation of all relevant agencies and institutions such as NRSC, NIC, National Mission on Sustainable Agriculture (NMSA),

MoRD (DoLR), SLUBs, AISLUS (All India Soil and Land Use Survey), SAUs and with ICAR (NBSSLUP) as mission leader.

The Development of National Portal on land and water Resources based on detailed Soil surveys at cadastral level, should be the priority of this group of institutions.

2. Concern was expressed on reliability, accuracy, timeliness, accessibility and effectiveness of land use data being generated by different organizations. Therefore, there is need for convergence of these databases on a common platform (scale and projection, etc.) using emerging spatial data/process standards to facilitate effective sharing, updation and utilization through NIC portal. There is a strong need to develop working linkage between organizations generating data and the user agencies by setting up a nodal agency at State, district and block levels and using them for development plans for providing the information at disaggregated levels.

    Action: State level RD&PR, Department of Agriculture, State Agriculture University (SAU) – extension units, KVKS, ETC, ATMA, NRSC, ATSLUS, ICAR.

22.3.2   ANALYZING SCIENTIFIC AND REMOTE SENSING DATA (MULTIPLE LAYER DATA ANALYSIS BY SCIENTIFIC AND RESEARCH INSTITUTIONS AT VARIOUS LEVELS)

1. The advances in remote sensing technologies have helped in generating rich and voluminous data on several parameters relating to land, water, forests, minerals, etc. These database need to be analyzed systematically not only to capture present land use practices but also to evolve optimal land use plans in conjunction with other resources. Such analysis of multilayer database (such as slope, soil type, soil fertility, rainfall pattern, moisture holding capacity of the soil, crop growth periods, etc.) help in assessing the resource potential and also in the development of perspective plans. The land not fit for agricultural and forest purposes can also be identified. Such lands can be allocated for industrial use especially for Special Economic Zones (SEZs) and other sectors.

2. The Remote sensing products, Geo-informatics tools and GIS platform may be fully internalized in developing both State level perspective land use plan and pragmatic farm level land use and management plan at village/watershed levels.

    Action: ICAR, NRSC, NIC, National Mission on Sustainable Agriculture (NMSA), MoRD (DoLR), MoAC, SLUBs, AISLUS (All India Soil and Land Use Survey)

3. Lessons learnt in the form of both success stories and failures from the ICAR – National Agriculture Technology Project (NATP) project – Mis-

sion Mode Project on Land Use Planning for Management of Agricultural Resources conducted so far may be shared with the local extension systems and development departments for ensuring follow up action and adoption.
Action: ICAR, SAUs, KVKS, ATMA, SLDs (State Line Departments)

4. A perspective land use policy and a Framework for location specific land use plan with SWOT analysis may be evolved for each state, for the efficient, sustainable management and utilization of the Natural Resources for realizing accelerated rural development and prosperity. The ICAR may take the lead role in coordinating and developing such a plan in collaboration with relevant R&D institutions and development departments in the states and the Centre.
   Action: ICAR, NRSC, NIC, NMSA, MoAC, MoRD (DoLR), SLUBs (State Land Use Boards), AISLUS.

22.3.3 POLICY IMPLICATION OF SCIENTIFIC DATABASES ON OPTIMUM LAND USE PLAN

1. It is evident that the current land use and also crop patterns and allied activities are largely determined/influenced by market forces. States have to study the implications of an optimal land use plan and evolve policies and strategies to regulate the market distortions, provide incentives to farmers through marketing arrangements, processing facilities and price supports. States should also sensitize officials to inculcate scientific temper among farmers. Such incessant efforts would facilitate in achieving the target, that is, agro-climatic specific optimal land use plan. The greening rural economy is integral part of such land use policies where in land allocations for forests, crops, fisheries, etc., will be guided by the rainfall, soil suitability, moisture holding capacity, productivity considerations. The systematic study of databases on natural resources should aid formulation of coherent policies for optimal use of the resources in an ecofriendly manner.

2. Weaning away increased use of marginal lands for agricultural production and arresting diversion of good agricultural lands for nonagricultural purposes may form part of land use policy implementation at the state level. Proper urban area planning could reduce the extent of conversion of good agricultural lands to nonagricultural uses. States can demarcate the lands for SEZs, industrial parks, housing, etc., and dovetail them into country and Town Planning Department Master plans.
   Action: NRSC, NIC, National Mission on Sustainable Agriculture (NMSA), MoRD (DoLR), SLUBs, AISLUS, ICAR & State Line Departments.

3. Institutional units and Joint operational mechanisms may be innovated/initiated/improvised for efficient managements of Natural resources (land, water, crops and livestock) for agriculture and Rural Development.

Action: Major departments of Government ICAR, SAUs, KVKs, State Line Departments, ATMA.

4. In both the underground and open cast mining, it is necessary to develop reclamation and rehabilitation strategies for the development of the post-mining land use in collaboration with the local population. Mining is a temporary land use and therefore progressive rehabilitation practices should be used to enable the land to return to a productive use as soon as possible after mining.

   Action: MoEF, NRSC, National Mission on Sustainable Agriculture (NMSA), MoRD (DoLR), SLUBs, Government Line Departments.

5. To ensure livelihoods of forest dwellers while protecting bio-diversity, forest based livelihoods have to be developed under state rural livelihood mission jointly with forest department and community, for example, Medicinal and Aromatic Plants, Bee Keeping, Tendu leaf collection, Non Timber Forest Produces (NTFPS).

   Action: MoRD (DoLR), MoA, MoWR, MoEF, MoTW&SJ.

6. Land use policy should be based on soil fertility – water resources and climate changes and land use and cropping pattern should be specified for each agro-ecological areas. Incentives should be provided for small and marginal farmers to reallocate their land for different crop and noncrop use in the context of such plans.

   Action: State Department of Agriculture, department of Water Resources and Irrigation, ICAR, SAU extension wings, KVKS, ETCs, ATMA, farmers' schools.

22.3.4. LEGAL FRAMEWORK

1. Land use, which was once perceived as a matter of local concern has become a national concern in the Globalization era and is being impinged upon by International Trade Agreements. Therefore, there is a need to amend the Constitution transferring the subject of land from State list to Concurrent list. Legal opinions may be sought and how is it possible under the constitution and the process to be followed, to ensure meeting the national targets of production of Agricultural commodities, their distribution and trade and to provide the Food and Nutritional Security to the people, now and in the future.

2. Viable land use planning and its implementation is closely linked to framing new land reforms instruments in the area of land tenure, land rights, land title and land lease deeds. These may be implemented by the concerned departments by enacting appropriate legislative measures.

   Action: MoRD (DoLR), Ministry of Law and Justice, State Line Departments.

22.3.5 AWARENESS AND MOBILIZING FARMER COMMUNITY

1. Awareness, Education, field visit, training and demonstration programs may be strengthened for the community for effective implementation of land use plans at different levels. Coastal Zone Management guidelines of the Ministry of Environment and Forest and implementation of Ecosystem-specific on-farm and nonfarm land based activities and land management programs, should be given wide publicity.
   Action: SAU extension wings, KVKS, ETC, ATMA, mobile extension vans, NIC web portals, State Line Departments

2. Information Communication and Technology (ICT) aided Decision support systems may be made available to the implementing agencies and user community levels for ecosystem specific land use and land management strategies and activities. This will promote participatory management and efficient conservation and utilization of the Natural Resources for Sustainable Agriculture.
   Action: State Line Departments, SAU extension wing, KVKS, ETC, ATMA, water users associations, Farmers associations, District level NIC centers, Ministry of Communication & IT at Centre and States.

3. Institutions and programs for capacity building may be strengthened at district and block levels including appropriate and Joint Action (Operational) Research and field demonstration programs for land use and water management programs in irrigated and rainfed farming systems.
   Action: SAU Extension units, ICAR institutions, ATMA, KVKs, reputed NGOs, State Line Departments.

4. Develop trained paraprofessionals viz., rural youth, progressive farmers, Panchayat elected representatives, women SHGs, labor banks at Block/Panchayat level to disseminate the knowledge through computer based tutorials at farmers school/Rytha Sampark Kendra or mobile extension vans with audio visual aids.
   Action: KVKS, ETC, ATMA, farmer schools, State Line Departments.

5. Capacity Building of the Block/ District level Agriculture officers, rural development officers, Fisheries officers, Livestock Officers, numbering about 25000, may be executed through training programs during the 12th Plan period. R & D institutions may develop "Quality training materials and case studies." This capacity building program may be of 2 weeks duration.
   Action: State Line Departments, SAU- extension wing, KVKs, ETC, ATMA, ICAR.

6. Horizontal coordination for convergence of concerned line departments' personnel and subject matter specialists from Research, Training and Education Institutions may be ensured during the planning, capacity building and monitoring of optimal land use plans.
   Action: State Line departments, SAU extension wing, KVKS, ETC, ATMA

22.4   NETWORK OF TRAINING INSTITUTIONS

1.   The efforts made in building the capacities of the grass roots beneficiaries by the KVK, SAU extension wing, ATMA, Agriculture department are not adequate. It is essential to link the institutions available at community level/ Gram Panchayat level, – The Water Users Association, JFM committees, SHGs, progressive farmers, reputed CBOs, NGOs, paraprofessional rural youth, etc., to reach the large mass of farming community with static and traveling visit and training programs.
     Action: State Line Departments, SAU extension wing, KVKS, ETC, CBOs, NGOs, ATMA, farmers associations, farmers schools.

22.5   CAPACITY BUILDING OF OFFICIALS—ELECTED REPRESENTATIVES AND FARMING COMMUNITIES

1.   Constitution of District level Training and Coordination Committee headed by the District Collector consisting of representatives of different line departments, Principal of ETC, etc., is necessary to guide planning, implementation and monitoring of land use plans.
2.   A Strong awareness generation and training programs are needed for the dissemination of the potential value of data/information generated from different sources. In order to achieve the desired result in short time, adoption of cascading mode of training with the help of other training institutions such as banking institutions, Panchayat Institutions, reputed CBOs, NGOs is recommended. It is necessary to develop the master trainers at Block and Gram panchayat level to disseminate the land use knowledge effectively through the progressive farmers, paraprofessionals rural educated youth and Women SHGs with land health care websites created.
     Action: Agriculture department, SAU extension wing, KVKS, ETC, ATMA, farmers associations, farmers schools, ICAR.

22.6   SUPPLY OF SCIENTIFIC DATABASES FOR EACH GRAM PANCHAYAT/VILLAGE

1.   Display of local Natural Resources and land use pattern at Panchayat Offices, may be promoted to help in local land health care by all stakeholders.
2.   Land use data, under the nine-fold classification, at more disaggregated levels may be made available to all the states, districts, blocks and villages.
     Action: District level NIC centers, Local technical institutions (Agriculture, watershed, horticulture, forestry, fisheries), Geography Departments of Colleges.

22.7   FACILITATING LAND USE PLANNING AND COMPREHENSIVE PLANS

1. Greening Economy – National policies advise for 1/3rd of the geographical area under green cover. Suitable areas may be identified by application of GIS tools, where farm/social forestry can be developed so that requirements of forest dependent population as well as consumers can be taken care off.
    Action: MoRD (DoLR), MoA, MoEF, MoTA, State RD&PR.
2. Agro-meteorology Advisory services in the form of Automatic Weather Station Installations, Rain Gauges at villages and training of climate risk management committees at village level may be strengthened to provide the first hand information on weather forecasting and action plan for a particular agro-climatic location and situation.
    Action: MoAC, IMD, NIC, National Mission on Sustainable Agriculture (NMSA), CRIDA under NICRA Project and State Line Departments.
3. Resource Planning, monitoring during the implementation phase and Post-facto social auditing of all the Agriculture and Rural Development programs may be undertaken at watershed/district level by a consortium of the members representing government, R&D institutions, credit supply agencies, NGOs and Gram Panchayat committees in an iterative process and these may be reviewed by the Gram Sabha of the Panchayat and by District Collector at district level.
    Action: Farmers, agriculture and water users associations, SAU, KVKS, ETC, ATMA, CBOs/NGOs; GP and District Planning and Coordination Committees.
4. Investment may be ear-marked for development of community – centered support services/structures under the Watershed Development Fund, so that the benefits continue to accrue even after the completion of the development program and the structures are maintained on a continuing basis.
    Action: State Level Nodal Agency, State Line Departments, Water Users Association, Gram Panchayat.
5. The associated support services in the form of community owned seed banks/fodder banks may be strengthened for adoption of contingency crop plans for coping with weather aberrations in rainfed areas.
    Action: SAU extension wing, Agriculture department, KVKS, Agriculture Research Stations, ATMA, Seed multiplication program through progressive farmers
6. Weather insurance protocols for rainfed crops may be standardized and operationalized at community level in the rainfed farming and drought prone areas.
    Action: Cooperative Grammeena and Nationalized banks, Agriculture department, Lombard – ICICI, NAIC.

7. ICT enabled remunerative agricultural marketing services as being promoted by the National Information Centre (NIC) through Agmarknet Nodes may be expanded in the country in diverse production zones.
    Action: NIC, SAU, KVKS, ATMA, RSETI.

8. Information Service on "Integrated Land Use Planning for Sustainable Agriculture," is to be introduced as touch screen SERVICE under the Agriculture Mission Mode project (AMMP), launched under the National e-Governance Program (NeGP), in collaboration with NBSS&LUP, SAUs and State Land Use Boards. Further, the users need to be trained to access the information available on touch screen.
    Action: ICAR, SAU- extension wing, KVKS, ETC, ATMA, NIC, State Line Departments.

9. Forward and backward linkages may be strengthened during the development of Special Economic Zones (SEZs) and related Special Agriculture Zones (SAZs) like Rainfed Crops Agro-Economic Zones (RAEZs), specific agriculture commodities production tracts/belts and peri-urban areas to ensure balanced growth of the associated sectors.

10. Contingency plans of Ministry of Agriculture should be integral while developing land use plans at district to village level
    Action: IMD, Agriculture Department, SAU, KVKS, ATMA, ICAR.

22.8   CONVERGENCE OF RURAL DEVELOPMENT AND AGRICULTURE SCHEMES AND THEIR MONITORING

1. There are several centrally sponsored and state specific land development programs such as seed improvement program, productivity enhancement program, promotion of organic farming, dry land horticulture, NHM, RKVY, MGNREGS, Minor irrigation department programs, etc. These need to be coordinated at the State level. It is felt that the departments are implementing the programs in isolation. Convergence of various schemes at village and beneficiary level would lead to better outcome than in isolation. It would be desirable if project specific ear-marked funding is continued to be provided under central sector by the GoI, so that the states will be implementing the programs without diversion of funds between programs.
    Action: MoRD (DoLR), MoA, MoWR, MoEF, MoTW, State RD&PR, MoF.

2. There is a need to constitute a State Level Training and Monitoring Co-ordination Committee. Nodal officers need to be identified at each level to monitor the progress of the various land and water based development programs. Similar committees may also be constituted at district, block and community level. The committees should take review of the progress of these programs, monitor and take corrective actions on a quarterly basis.

Action: State and District level coordination Committees

3. The forum has also suggested to carryout few research studies on topics such as:

   a. Land use/Land cover changes in India: A Design for Monitoring Networking

   b. Integrated Module to predict Indian Land use – climate module, Land degradation module, soil and crop module, Hydrological module, land use and management and development module for rural development, regional impact assessment module.

KEYWORDS

- **Information Communication and Technology**
- **Land Resource Maps**
- **Land Use Management**
- **Land Use Plan**
- **Legal Framework**
- **National Agriculture Technology Project**
- **National Mission on Sustainable Agriculture**
- **Remote Sensing Data**
- **Special Economic Zones**

INDEX